www.wadsworth.com

wadsworth.com is the World Wide Web site for Wadsworth and is your direct source to dozens of online resources.

At *wadsworth.com* you can find out about supplements, demonstration software, and student resources. You can also send email to many of our authors and preview new publications and exciting new technologies.

wadsworth.com
Changing the way the world learns®

Classic Readings in Sociology

Third Edition

EVE L. HOWARD

THOMSON

™

WADSWORTH

Australia • Canada • Mexico • Singapore • Spain
United Kingdom • United States

Sociology Editor: *Bob Jucha*
Assistant Editor: *Stephanie Monzon*
Editorial Assistant: *Melissa Walter*
Technology Project Manager: *Dee Dee Zobian*
Marketing Manager: *Matthew Wright*
Marketing Assistant: *Michel Silverstein*
Advertising Project Manager: *Linda Yip*
Project Manager, Editorial Production:
    Emily Smith

Print/Media Buyer: *Rebecca Cross*
Permissions Editor: *Kiely Sexton*
Production Service and Compositor: *Buuji, Inc.*
Copy Editor: *Alan DeNiro, Buuji, Inc.*
Cover Designer: *Howard Burch*
Cover Image: © *Hulton Picture Library*
Text and Cover Printer: *Webcom*

For more information about our products,
contact us at:
Thomson Learning Academic Resource Center
1-800-423-0563

For permission to use material from this text,
contact us by:
Phone: 1-800-730-2214   Fax: 1-800-730-2215
Web: http://www.thomsonrights.com

Library of Congress Control Number:
    2003104456

ISBN 0-534-60975-9

Wadsworth/Thomson Learning
10 Davis Drive
Belmont, CA 94002-3098
USA

Asia
Thomson Learning
5 Shenton Way #01-01
UIC Building
Singapore 068808

Australia/New Zealand
Thomson Learning
102 Dodds Street
Southbank, Victoria 3006
Australia

Canada
Nelson
1120 Birchmount Road
Toronto, Ontario M1K 5G4
Canada

Europe/Middle East/Africa
Thomson Learning
High Holborn House
50/51 Bedford Row
London WC1R 4LR
United Kingdom

Latin America
Thomson Learning
Seneca, 53
Colonia Polanco
11560 Mexico D.F.
Mexico

Spain/Portugal
Paraninfo
Calle/Magallanes, 25
28015 Madrid, Spain

Contents

Preface

Classic Readings in Sociology has been carefully designed to include selections that teachers of introductory sociology and theory courses most often want their students to read. The articles were selected to represent past and present sociological thought that has endured and will endure to become classic in the field. The low price of this collection makes it an ideal accompaniment to a standard textbook—or, for instructors who don't use a text, this reader (perhaps used with a selection of other books) gives students a strong introduction to the foundation of the field.

The articles were chosen after reviewing the results of an extensive survey of instructors' input of timeless selections they use in their teaching of Sociology. Included are excerpts from timeless, well-known works such as C. Wright Mills's "The Promise of Sociology" and Peter Berger's "Invitation to Sociology," as well as selections from highly regarded contemporary writings such as Jonathan Kozol's "Savage Inequalities." These works help students develop a sense of the nature of the discipline and its history, exposing them to original thinking and analysis.

The collection includes readings that can accompany each part of an introductory course. For instance, Peter Berger's "Invitation to Sociology" provides a perfect way to introduce the nature of sociology to students. "The Self" by George Herbert Mead, can easily supplement a discussion of socialization; an excerpt from Mills's "The Power Elite" can augment work on the political world. The articles can promote stimulating classroom discussions and form the basis for engaging writing assignments. Most important, the articles provide the tools to help students understand how sociologists think, and they lay the foundation for future reading and learning.

Classic Readings in Sociology includes several additional helpful elements. A full glossary of sociological terms will especially benefit instructors who choose to use this collection without an accompanying textbook. Web site links are provided to help students and instructors access recent data and information. These links can provide another way to prompt classroom discussion, connect the classic articles to more recent writings, and supplement research for a written paper or exercise.

What is new in the third edition? The first and second editions enjoyed resoundingly positive feedback. The selections have been validated as truly seminal works in the field. In reviewing for the third edition we were asked to add discussion questions after each work to enable students and professors to integrate the thought and discussion of the original work into a relevant conceptual framework complementing course goals. You will find three to four questions that ask students to think critically, perform an activity to increase understanding, and perhaps link a thought from one article to the article at hand. The discussion questions build in difficulty and sophistication, applicable to students in an upper-division theory course.

A new article has been added as well, by David L. Rosenhan, "Being Sane in Insane Places." This article focuses on whether we can understand the differences between the label of insane and sane, which has great applicability to the understanding of basic tenets of the concept of Deviance and Socialization.

Enjoy the collection and please email me with ideas for improvement.

Eve L. Howard
Editor in Chief, Social Sciences
eve.howard@wadsworth.com

1

The Promise of Sociology

C. WRIGHT MILLS

Nowadays men often feel that their private lives are a series of traps. They sense that within their everyday worlds, they cannot overcome their troubles, and in this feeling, they are often quite correct: What ordinary men are directly aware of and what they try to do are bounded by the private orbits in which they live; their visions and their powers are limited to the close-up scenes of job, family, neighborhood; in other milieux, they move vicariously and remain spectators. And the more aware they become, however vaguely, of ambitions and of threats which transcend their immediate locales, the more trapped they seem to feel.

Underlying this sense of being trapped are seemingly impersonal changes in the very structure of continent-wide societies. The facts of contemporary history are also facts about the success and the failure of individual men and women. When a society is industrialized, a peasant becomes a worker; a feudal lord is liquidated or becomes a businessman. When classes rise or fall, a man is employed or unemployed; when the rate of investment goes up or down, a man takes new heart or goes broke. When wars happen, an insurance salesman becomes a rocket launcher; a

store clerk, a radar man; a wife lives alone; a child grows up without a father. Neither the life of an individual nor the history of a society can be understood without understanding both.

Yet men do not usually define the troubles they endure in terms of historical change and institutional contradiction. The well-being they enjoy, they do not usually impute to the big ups and downs of the societies in which they live. Seldom aware of the intricate connection between the patterns of their own lives and the course of world history, ordinary men do not usually know what this connection means for the kinds of men they are becoming and for the kinds of history-making in which they might take part. They do not possess the quality of mind essential to grasp the interplay of man and society, or biography and history, of self and world. They cannot cope with their personal troubles in such ways as to control the structural transformations that usually lie behind them.

Surely it is no wonder. In what period have so many men been so totally exposed at so fast a pace to such earthquakes of change? That Americans have not known such catastrophic changes as have the men and women of other societies is due to

Credit: From *The Sociology Imagination* by C. Wright Mills, © 2000 by Oxford University Press, Inc. Used by pemission of Oxford University Press, Inc.

historical facts that are now quickly becoming "merely history." The history that now affects every man is world history. Within this scene and this period, in the course of a single generation, one-sixth of mankind is transformed from all that is feudal and backward into all that is modern, advanced, and fearful. Political colonies are freed; new and less visible forms of imperialism installed. Revolutions occur; men feel the intimate grip of new kinds of authority. Totalitarian societies rise, and are smashed to bits—or succeed fabulously. After two centuries of ascendancy, capitalism is shown up as only one way to make society into an industrial apparatus. After two centuries of hope, even formal democracy is restricted to a quite small portion of mankind. Everywhere in the underdeveloped world, ancient ways of life are broken up and vague expectations become urgent demands. Everywhere in the overdeveloped world, the means of authority and of violence become total in scope and bureaucratic in form. Humanity itself now lies before us, the super-nation at either pole concentrating its most coordinated and massive efforts upon the preparation of World War Three.

The very shaping of history now outpaces the ability of men to orient themselves in accordance with cherished values. And which values? Even when they do not panic, men often sense that older ways of feeling and thinking have collapsed and that newer beginnings are ambiguous to the point of moral stasis. Is it any wonder that ordinary men feel they cannot cope with the larger worlds with which they are so suddenly confronted? That they cannot understand the meaning of their epoch for their own lives? That—in defense of selfhood—they become morally insensible, trying to remain altogether private men? Is it any wonder that they come to be possessed by a sense of the trap?

It is not only information that they need—in this Age of Fact, information often dominates their attention and overwhelms their capacities to assimilate it. It is not only the skills of reason that they need—although their struggles to acquire these often exhaust their limited moral energy.

What they need, and what they feel they need, is a quality of mind that will help them to use information and to develop reason in order to achieve lucid summations of what is going on in the world and of what may be happening within themselves. It is this quality, I am going to contend, that journalists and scholars, artists and publics, scientists and editors are coming to expect of what may be called the sociological imagination.

The sociological imagination enables its possessor to understand the larger historical scene in terms of its meaning for the inner life and the external career of a variety of individuals. It enables him to take into account how individuals, in the welter of their daily experience, often become falsely conscious of their social positions. Within that welter, the framework of modern society is sought, and within that framework the psychologies of a variety of men and women are formulated. By such means the personal uneasiness of individuals is focused upon explicit troubles and the indifference of publics is transformed into involvement with public issues.

The first fruit of this imagination—and the first lesson of the social science that embodies it—is the idea that the individual can understand his own experience and gauge his own fate only by locating himself within his period, that he can know his own chances in life by becoming aware of those of all individuals in his circumstances. In many ways it is a terrible lesson; in many ways a magnificent one. We do not know the limits of man's capacities for supreme effort or willing degradation, for agony or glee, for pleasurable brutality or the sweetness of reason. But in our time we have come to know that the limits of "human nature" are frighteningly broad. We have come to know that every individual lives, from one generation to the next, in some society; that he lives out a biography, and that he lives it out within some historical sequence. By the fact of his living he contributes, however minutely, to the shaping of this society and to the course of its history, even as he is made by society and by its historical push and shove.

The sociological imagination enables us to grasp history and biography and the relations between the two within society. That is its task and its promise. To recognize this task and this promise is the mark of the classic social analyst. It is characteristic of Herbert Spencer—turgid, polysyllabic, comprehensive; of E. A. Ross—graceful, muckraking, upright; of August Comte and Emile Durkheim; of the intricate and subtle Karl Mannheim. It is the quality of all that is intellectually excellent in Karl Marx; it is the clue to Thorstein Veblen's brilliant and ironic insight, to Joseph Schumpeter's many-sided constructions of reality; it is the basis of the psychological sweep of W. E. H. Lecky no less than of the profundity and clarity of Max Weber. And it is the signal of what is best in contemporary studies of man and society.

No social study that does not come back to the problems of biography, of history, and of their intersections within a society has completed its intellectual journey. Whatever the specific problems of the classic social analysts, however limited or however broad the features of social reality they have examined, those who have been imaginatively aware of the promise of their work have consistently asked three sorts of questions:

(1) What is the structure of this particular society as a whole? What are its essential components, and how are they related to one another? How does it differ from other varieties of social order? Within it, what is the meaning of any particular feature for its continuance and for its change?

(2) Where does this society stand in human history? What are the mechanics by which it is changing? What is its place within and its meaning for the development of humanity as a whole? How does any particular feature we are examining affect, and how is it affected by, the historical period in which it moves? And this period—what are its essential features? How does it differ from other periods? What are its characteristic ways of history-making?

(3) What varieties of men and women now prevail in this society and in this period? And what varieties are coming to prevail? In what ways are they selected and formed, liberated and repressed, made sensitive and blunted? What kinds of "human nature" are revealed in the conduct and character we observe in this society in this period? And what is the meaning for "human nature" of each and every feature of the society we are examining?

Whether the point of interest is a great power state or a minor literary mood, a family, a prison, a creed—these are the kinds of questions the best social analysts have asked. They are the intellectual pivots of classic studies of man in society—and they are questions inevitably raised by any mind possessing the sociological imagination. For that imagination is the capacity to shift from one perspective to another—from the political to the psychological; from examination of a single family to comparative assessment of the national budgets of the world; from the theological school to the military establishment; from considerations of an oil industry to studies of contemporary poetry. It is the capacity to range from the most impersonal and remote transformations to the most intimate features of the human self—and to see the relations between the two. Back of its use there is always the urge to know the social and historical meaning of the individual in the society and in the period in which he has his quality and his being.

That, in brief, is why it is by means of the sociological imagination that men now hope to grasp what is going on in the world, and to understand what is happening in themselves as minute points of the intersections of biography and history within society. In large part, contemporary man's self-conscious view of himself as at least an outsider, if not a permanent stranger, rests upon an absorbed realization of social relativity and of the transformative power of history. The sociological imagination is the most fruitful form of this self-consciousness. By its use men whose mentalities have swept only a series of limited orbits often come to feel as if suddenly awakened in a house with which they had only supposed themselves to be familiar.

Correctly or incorrectly, they often come to feel that they can now provide themselves with adequate summations, cohesive assessments, comprehensive orientations. Older decisions that once appeared sound now seem to them products of a mind unaccountably dense. Their capacity for astonishment is made lively again. They acquire a new way of thinking, they experience a transvaluation of values: In a word, by their reflection and by their sensibility, they realize the cultural meaning of the social sciences.

Perhaps the most fruitful distinction with which the sociological imagination works is between "the personal troubles of milieu" and "the public issues of social structure." This distinction is an essential tool of the sociological imagination and a feature of all classic work in social science.

Troubles occur within the character of the individual and within the range of his immediate relations with others; they have to do with his self and with those limited areas of social life of which he is directly and personally aware. Accordingly, the statement and the resolution of troubles properly lie within the individual as a biographical entity and within the scope of this immediate milieu—the social setting that is directly open to his personal experience and to some extent his willful activity. A trouble is a private matter: Values cherished by an individual are felt by him to be threatened.

Issues have to do with matters that transcend these local environments of the individual and the range of his inner life. They have to do with the organization of many such milieux into the institutions of an historical society as a whole, with the ways in which various milieux overlap and interpenetrate to form the larger structure of social and historical life. An issue is a public matter: Some value cherished by publics is felt to be threatened. Often there is a debate about what that value really is and about what it is that really threatens it. This debate is often without focus if only because it is the very nature of an issue, unlike even widespread trouble, that it cannot

very well be defined in terms of the immediate and everyday environments of ordinary men. An issue, in fact, often involves a crisis in institutional arrangements, and often too it involves what Marxists call "contradictions" or "antagonisms."

In these terms, consider unemployment. When, in a city of 100,000, only one man is unemployed, that is his personal trouble, and for its relief we properly look to the character of the man, his skills, and his immediate opportunities. But when in a nation of 50 million employees, 15 million men are unemployed, that is an issue, and we may not hope to find its solution within the range of opportunities open to any one individual. The very structure of opportunities has collapsed. Both the correct statement of the problem and the range of possible solutions require us to consider the economic and political institutions of the society, and not merely the personal situation and character of a scatter of individuals.

Consider war. The personal problem of war, when it occurs, may be how to survive it or how to die in it with honor; how to make money out of it; how to climb into the higher safety of the military apparatus; or how to contribute to the war's termination. In short, according to one's values, to find a set of milieux and within it to survive the war or make one's death in it meaningful. But the structural issues of war have to do with its causes; with what types of men it throws up into command; with its effects upon economic and political, family and religious institutions, with the unorganized irresponsibility of a world of nation-states.

Consider marriage. Inside a marriage a man and a woman may experience personal troubles, but when the divorce rate during the first four years of marriage is 250 out of every 1,000 attempts, this is an indication of a structural issue having to do with the institutions of marriage and the family and other institutions that bear upon them.

Or consider the metropolis—the horrible, beautiful, ugly, magnificent sprawl of the great city. For many upper-class people, the personal solution to "the problem of the city" is to have an apartment with private garage under it in the heart of the city and, forty miles out, a house by Henry Hill, garden by Garrett Eckbo, on a hundred acres of private land. In these two controlled environments—with a small staff at each end and a private helicopter connection—most people could solve many of the problems of personal milieux caused by the facts of the city. But all this, however splendid, does not solve the public issues that the structural fact of the city poses. What should be done with this wonderful monstrosity? Break it up into scattered units, combining residence and work? Refurbish it as it stands? Or, after evacuation, dynamite it and build new cities according to new plans in new places? What should those plans be? And who is to decide and to accomplish whatever choice is made? These are structural issues; to confront them and to solve them requires us to consider political and economic issues that affect innumerable milieu.

Insofar as an economy is so arranged that slumps occur, the problem of unemployment becomes incapable of personal solution. Insofar as war is inherent in the nation-state system and in the uneven industrialization of the world, the ordinary individual in his restricted milieu will be powerless—with or without psychiatric aid—to solve the troubles this system or lack of system imposes upon him. Insofar as the family as an institution turns women into darling little slaves and men into their chief providers and unweaned dependents, the problem of a satisfactory marriage remains incapable of purely private solution. Insofar as the overdeveloped megalopolis and the overdeveloped automobile are built-in features of the overdeveloped society, the issues of urban living will not be solved by personal ingenuity and private wealth.

What we experience in various and specific milieux, I have noted, is often caused by structural changes. Accordingly, to understand the changes of many personal milieux we are required to look beyond them. And the number and variety of such structural changes increase as the institutions within which we live become more embracing and more intricately connected with one another. To be aware of the idea of social structure and to use it with sensibility is to be capable of tracing such linkages among a great variety of milieux. To be able to do that is to possess the sociological imagination. . . .

DISCUSSION QUESTIONS

1. Mills argues that personal troubles can be understood in terms of large-scale patterns that extend beyond individual experience and are part of society and history. Identify an issue that you think would be a personal trouble for an individual, and then identify the societal influences that you think impinge on the individual's experience. (Hint: you might think of such things as divorce, violence, or school failure.)

2. C. Wright Mills identifies the central task of sociology to be grasping the relationships between history and biography. To do this, identify two large-scale historical events that you think have most shaped your biography.

Now ask someone of a different generation that same question. What does this tell you about how sociologists think about the relationship between individuals and society?

3. Sociologists sometimes distinguish between micro-level and macro-level theories. Both study social influences on human lives. Take a topic (such as terrorism, divorce, or illness) and, using the perspective that Mills articulates, how might someone using a micro-theoretical approach study this topic? How would that differ from how someone using a macro-level theoretical approach would study the same topic?

INTERNET RESOURCES

Suggested Web URLs
for Further Study

http://www.ac.wwu.edu/~stephan/timeline.html
This Web site contains a basic timeline of important people and papers in sociology beginning from its basis in philosophy.

http://www.socioweb.com/~markbl/socioweb/
Independent guide to Sociological Resources on the web.

InfoTrac College Edition

You can find further relevant readings on the World Wide Web at
http://sociology.wadsworth.com

Virtual Society

For further information on this subject including links to relevant Web sites, go to the Wadsworth Sociology homepage at
http://sociology.wadsworth.com

2

Invitation to Sociology

PETER L. BERGER

It can be said that the first wisdom of sociology is this—things are not what they seem. This too is a deceptively simple statement. It ceases to be simple after a while. Social reality turns out to have many layers of meaning. The discovery of each new layer changes the perception of the whole.

Anthropologists use the term "culture shock" to describe the impact of a totally new culture upon a newcomer. In an extreme instance such shock will be experienced by the Western explorer who is told, halfway through dinner, that he is eating the nice old lady he had been chatting with the previous day—a shock with predictable physiological if not moral consequences. Most explorers no longer encounter cannibalism in their travels today. However, the first encounters with polygamy or with puberty rites or even with the way some nations drive their automobiles can be quite a shock to an American visitor. With the shock may go not only disapproval or disgust but a sense of excitement that things can really be that different from what they are at home. To some extent, at least, this is the excitement of any first travel abroad. The experience of sociological discovery could be described as "culture shock" minus geographical displacement. In other words, the sociologist travels at home—with shocking results. He is unlikely to find that he is eating a nice old lady for dinner. But the discovery, for instance, that his own church has considerable money invested in the missile industry or that a few blocks from his home there are people who engage in cultic orgies may not be drastically different in emotional impact. Yet we would not want to imply that sociological discoveries are always or even usually outrageous to moral sentiment. Not at all. What they have in common with exploration in distant lands, however, is the sudden illumination of new and unsuspected facets of human existence in society. This is the excitement and, as we shall try to show later, the humanistic justification of sociology.

People who like to avoid shocking discoveries, who prefer to believe that society is just what they were taught in Sunday school, who like the safety of the rules and the maxims of what Alfred Schuetz has called the "world-taken-for-granted," should stay away from sociology. People who feel no temptation before closed doors, who have no curiosity about human beings, who are

Credit: From *An Invitation to Sociology* by Peter Berger, © 1963 by Peter L. Berger. Used by permission of Doubleday, a division of Random House, Inc.

content to admire scenery without wondering about the people who live in those houses on the other side of that river, should probably also stay away from sociology. They will find it unpleasant or, at any rate, unrewarding. People who are interested in human beings only if they can change, convert, or reform them should also be warned, for they will find sociology much less useful than they hoped. And people whose interest is mainly in their own conceptual constructions will do just as well to turn to the study of little white mice. Sociology will be satisfying, in the long run, only to those who can think of nothing more entrancing than to watch men and to understand things human. . . .

To ask sociological questions, then, presupposes that one is interested in looking some distance beyond the commonly accepted or officially defined goals of human actions. It presupposes a certain awareness that human events have different levels of meaning, some of which are hidden from the consciousness of everyday life. It may even presuppose a measure of suspicion about the way in which human events are officially interpreted by the authorities, be they political, juridical, or religious in character. If one is willing to go as far as that, it would seem evident that not all historical circumstances are equally favorable for the development of sociological perspective.

It would appear plausible, in consequence, that sociological thought would have the best chance to develop in historical circumstances marked by severe jolts to the self-conception, especially the official and authoritative and generally accepted self-conception of a culture. It is only in such circumstances that perceptive men are likely to be motivated to think beyond the assertions of this self-conception and, as a result, question the authorities. . . .

Sociological perspective can then be understood in terms of such phrases as "seeing through," "looking behind," very much as such phrases would be employed in common speech—

"seeing through his game," "looking behind the scenes"—in other words, "being up on all the tricks."

. . . We could think of this in terms of a common experience of people living in large cities. One of the fascinations of a large city is the immense variety of human activities taking place behind the seemingly anonymous and endlessly undifferentiated rows of houses. A person who lives in such a city will time and again experience surprise or even shock as he discovers the strange pursuits that some men engage in quite unobtrusively in houses that, from the outside, look like all the others on a certain street. Having had this experience once or twice, one will repeatedly find oneself walking down a street, perhaps late in the evening, and wondering what may be going on under the bright lights showing through a line of drawn curtains. An ordinary family engaged in pleasant talk with guests? A scene of desperation amid illness or death? Or a scene of debauched pleasures? Perhaps a strange cult or a dangerous conspiracy? The facades of the houses cannot tell us, proclaiming nothing but an architectural conformity to the tastes of some group or class that may not even inhabit the street any longer. The social mysteries lie behind the facades. The wish to penetrate to these mysteries is an analogon to sociological curiosity. In some cities that are suddenly struck by calamity this wish may be abruptly realized. Those who have experienced wartime bombings know of the sudden encounters with unsuspected (and sometimes unimaginable) fellow tenants in the air-raid shelter of one's apartment building. Or they can recollect the startling morning sight of a house hit by a bomb during the night, neatly sliced in half, the facade torn away and the previously hidden interior mercilessly revealed in the daylight. But in most cities that one may normally live in, the facades must be penetrated by one's own inquisitive intrusions. Similarly, there are historical situations in which the facades of society are violently torn apart and all but the most incurious are forced to see that there was a reality behind the facades all along. Usually this does not happen, and the facades continue to confront us with seemingly rocklike permanence. The perception of the reality behind the facades then demands a considerable intellectual effort.

A few examples of the way in which sociology "looks behind" the facades of social structures might serve to make our argument clearer. Take, for instance, the political organization of a community. If one wants to find out how a modern American city is governed, it is very easy to get the official information about this subject. The city will have a charter, operating under the laws of the state. With some advice from informed individuals, one may look up various statutes that define the constitution of the city. Thus one may find out that this particular community has a city-manager form of administration, or that party affiliations do not appear on the ballot in municipal elections, or that the city government participates in a regional water district. In similar fashion, with the help of some newspaper reading, one may find out the officially recognized political problems of the community. One may read that the city plans to annex a certain suburban area, or that there has been a change in the zoning ordinances to facilitate industrial development in another area, or even that one of the members of the city council has been accused of using his office for personal gain. All such matters still occur on the, as it were, visible, official, or public level of political life. However, it would be an exceedingly naive person who would believe that this kind of information gives him a rounded picture of the political reality of that community. The sociologist will want to know above all the constituency of the "informal power structure" (as it has been called by Floyd Hunter, an American sociologist interested in such studies), which is a configuration of men and their power that cannot be found in any statutes, and probably cannot be read about in the newspapers. The

political scientist or the legal expert might find it very interesting to compare the city charter with the constitutions of other similar communities. The sociologist will be far more concerned with discovering the way in which powerful vested interests influence or even control the actions of officials elected under the charter. These vested interests will not be found in city hall, but rather in the executive suites of corporations that may not even be located in that community, in the private mansions of a handful of powerful men, perhaps in the offices of certain labor unions, or even, in some instances, in the headquarters of criminal organizations. When the sociologist concerns himself with power, he will "look behind" the official mechanisms that are supposed to regulate power in the community. This does not necessarily mean that he will regard the official mechanisms as totally ineffective or their legal definition as totally illusionary. But at the very least he will insist that there is another level of reality to be investigated in the particular system of power. In some cases he might conclude that to look for real power in the publicly recognized places is quite delusional. . . .

Let us take one further example. In Western countries, and especially in America, it is assumed that men and women marry because they are in love. There is a broadly based popular mythology about the character of love as a violent, irresistible emotion that strikes where it will, a mystery that is the goal of most young people and often of the not-so-young as well. As soon as one investigates, however, which people actually marry each other, one finds that the lightning-shaft of Cupid seems to be guided rather strongly within very definite channels of class, income, education, [and] racial and religious background. If one then investigates a little further into the behavior that is engaged in prior to marriage under the rather misleading euphemism of "courtship," one finds channels of interaction that are often rigid to the point of ritual. The suspicion begins to dawn on one that, most of the time, it is not so much the emotion of love that creates a certain kind of relationship, but that carefully predefined and often planned relationships eventually generate the desired emotion. In other words, when certain conditions are met or have been constructed, one allows oneself "to fall in love." The sociologist investigating our patterns of "courtship" and marriage soon discovers a complex web of motives related in many ways to the entire institutional structure within which an individual lives his life—class, career, economic ambition, aspirations of power and prestige. The miracle of love now begins to look somewhat synthetic. Again, this need not mean in any given instance that the sociologist will declare the romantic interpretation to be an illusion. But, once more, he will look beyond the immediately given and publicly approved interpretations. . . .

We would contend, then, that there is a debunking motif inherent in sociological consciousness. The sociologist will be driven time and again, by the very logic of his discipline, to debunk the social systems he is studying. This unmasking tendency need not necessarily be due to the sociologist's temperament or inclinations. Indeed, it may happen that the sociologist, who as an individual may be of a conciliatory disposition and quite disinclined to disturb the comfortable assumptions on which he rests his own social existence, is nevertheless compelled by what he is doing to fly in the face of what those around him take for granted. In other words, we would contend that the roots of the debunking motif in sociology are not psychological but methodological. The sociological frame of reference, with its built-in procedure of looking for levels of reality other than those given in the official interpretations of society, carries with it a logical imperative to unmask the pretensions and the propaganda by which men cloak their actions with each other. This unmasking imperative is one of the characteristics of sociology particularly at home in the temper of the modern era . . .

DISCUSSION QUESTIONS

1. Berger's argument makes the case for sociology as a debunking perspective—that is, looking behind the taken-for-granted ways that people usually see things. He uses the example of marriage by showing that, although people think of marriage as happening just because people are in love, there are actually quite specific social patterns about who marries whom. What social factors—not just romantic factors—do you think influence people's choices of marriage partners?

2. Berger suggests that some historical conditions may be more conducive than others to producing a sociological perspective—particularly if what he calls the "facades of society" are torn apart. Think of an example of such an event and discuss how this event may have challenged people's prior ways of thinking about society.

3. Implicit in Berger's argument is the idea that those who are content with the status quo may not be as able to acquire a sociological perspective. Why would this be true?

INTERNET RESOURCES

Suggested Web URLs
for Further Study

http://sociology.wadsworth.com
Society for Applied Sociology, founded in 1978, is an international organization for professionals involved in applying sociological knowledge in a wide variety of settings.

http://www.angelfire.com/or/sociologyshop/PLB.html
This Web site is dedicated to Peter L. Berger and his work, especially *Invitation to Sociology* and *The Social Construction of Reality.*

InfoTrac College Edition

You can find further relevant readings on the World Wide Web at
http://sociology.wadsworth.com

Virtual Society

For further information on this subject including links to relevant Web sites, go to the Wadsworth Sociology homepage at
http://sociology.wadsworth.com

3

Manifesto of the Communist Party

KARL MARX AND FRIEDRICH ENGELS

BOURGEOIS AND PROLETARIANS[1]

The history of all hitherto existing society[2] is the history of class struggles.

Freeman and slave, patrician and plebeian, lord and self, guild-master[3] and journeyman, in a word, oppressor and oppressed, stood in constant opposition to one another, carried on an uninterrupted, now hidden, now open fight, a fight that each time ended, either in a revolutionary reconstitution of society at large, or in the common ruin of the contending classes.

In the earlier epochs of history, we find almost everywhere a complicated arrangement of society into various orders, a manifold gradation of social rank. In ancient Rome we have patricians, knights, plebeians, slaves; in the Middle Ages, feudal lords, vassals, guild-masters, journeymen, apprentices, serfs; in almost all of these classes, again, subordinate gradations.

The modern bourgeois society that has sprouted from the ruins of feudal society, has not done away with class antagonisms. It has but established new classes, new conditions of oppression, new forms of struggle in place of the old ones.

Our epoch, the epoch of the bourgeoisie, possesses, however, this distinctive feature; it has simplified the class antagonisms. Society as a whole is more and more splitting up into two great hostile camps, into two great classes directly facing each other: Bourgeoisie and Proletariat.

From the serfs of the Middle Ages sprang the chartered burghers of the earliest towns. From these burgesses the first elements of the bourgeoisie were developed.

The discovery of America, the rounding of the Cape, opened up fresh ground for the rising bourgeoisie. The East Indian and Chinese markets, the [colonization] of America, trade with the colonies, the increase in the means of exchange and in commodities generally, gave to commerce, to navigation, to industry, an impulse never before known, and thereby, to the revolutionary element in the tottering feudal society, a rapid development.

The feudal system of industry, under which industrial production was monopolized by close guilds, now no longer sufficed for the growing wants of the new markets. The manufacturing system took its place. The guild-masters were pushed on one side by the manufacturing middle class; division of labor between the different corporate guilds vanished in the face of division of labor in each single workshop.

Meantime the markets kept ever growing, the demand, ever rising. Even manufacture no longer sufficed. Thereupon, steam and machinery revolutionized industrial production. The place of manufacture was taken by the giant, Modern

Industry, the place of the industrial middle class, by industrial millionaires, the leaders of whole industrial armies, the modern bourgeois.

Modern industry has established the world-market, for which the discovery of America paved the way. This market has given an immense development to commerce, to navigation, to communication by land. This development has, in its turn, reacted on the extension of industry; and in proportion as industry, commerce, navigation, railways extended, in the same proportion the bourgeoisie developed, increased its capital, and pushed into the background every class handed down from the Middle Ages.

We see, therefore, how the modern bourgeoisie is itself the product of a long course of development, of a series of revolutions in the modes of production and of exchange.

Each step in the development of the bourgeoisie was accompanied by a corresponding political advance of that class. An oppressed class under the sway of the feudal nobility, an armed and self-governing association in the mediaeval commune,[4] here independent urban republic (as in Italy and Germany), there taxable "third estate" of the monarchy (as in France), afterwards, in the period of manufacture proper, serving either the semifeudal or the absolute monarchy as a counterpoise against the nobility, and, in fact, cornerstone of the great monarchies in general, the bourgeoisie has at last, since the establishment of Modern Industry and of the world-market, conquered for itself, in the modern representative State, exclusive political sway. The executive of the modern State is but a committee for managing the common affairs of the whole bourgeoisie.

The bourgeoisie, historically, has played a most revolutionary part.

The bourgeoisie, wherever it has got the upper hand, has put an end to all feudal, patriarchal, idyllic relations. It has pitilessly torn asunder the motley feudal ties that bound man to his "natural superiors," and has left remaining no other nexus between man and man than naked self-interest, than callous "cash payment." It has drowned the most heavenly ecstasies of religious fervour, of chivalrous enthusiasm, of philistine sentimentalism, in the icy water of egotistical calculation. It has resolved personal worth into exchange value, and in place of the numberless indefeasible chartered freedoms, has set up that single, unconscionable freedom—Free Trade. In one word, for exploitation, veiled by religious and political illusions, it has substituted naked, shameless, direct, brutal exploitation.

The bourgeoisie has stripped of its halo every occupation hitherto honoured and looked up to with reverent awe. It has converted the physician, the lawyer, the priest, the poet, the man of science, into its paid [wage-laborers].

The bourgeoisie has torn away from the family its sentimental veil, and has reduced the family relation to a mere money relation.

The bourgeoisie has disclosed how it came to pass that the brutal display of vigour in the Middle Ages, which Reactionists so much admire, found its fitting complement in the most slothful indolence. It has been the first to show what man's activity can bring about. It has accomplished wonders far surpassing Egyptian pyramids, Roman aqueducts, and Gothic cathedrals; it has conducted expeditions that put in the shade all former Exoduses of nations and crusades.

The bourgeoisie cannot exist without constantly revolutionizing the instruments of production, and thereby the relations of production, and with them the whole relations of society. Conservation of the old modes of production in unaltered form, was, on the contrary, the first condition of existence for all earlier industrial classes. Constant revolutionizing of production, uninterrupted disturbance of all social conditions, everlasting uncertainty and agitation distinguish the bourgeois epoch from all earlier ones. All fixed, fast-frozen relations, which their train of ancient and venerable prejudices and opinions, are swept away, all new formed ones become antiquated before they can ossify. All that is solid melts into air, all that is holy is profaned, and man is at last compelled to face with sober senses, his real conditions of life, and his relations with his kind.

The need of a constantly expanding market for its products chases the bourgeoisie over the whole surface of the globe. It must nestle everywhere, settle everywhere, establish [connections] everywhere.

The bourgeoisie has through its exploitation of the world-market given a cosmopolitan character to production and consumption in every country. To the great chagrin of Reactionists, it has drawn from under the feet of industry the national ground on which it stood. All old-established national industries have been destroyed or are daily being destroyed. They are dislodged by new industries, whose introduction becomes a life and death question for all civilised nations, by industries that no longer work up indigenous raw material, but rare material drawn from the remotest zones; industries whose products are consumed, not only at home, but in every quarter of the globe. In place of the old wants, satisfied by the productions of the country, we find new wants, requiring for their satisfaction the products of distant lands and climes. In place of the old local and national seclusion and self-sufficiency, we have intercourse in every direction, universal interdependence of nations. And as in material, so also in intellectual production. The intellectual creations of individual nations become common property. National one-sidedness and narrow-mindedness become more and more impossible, and from the numerous national and local literatures there arises a world-literature.

The bourgeoisie, by the rapid improvement of all instruments of production, by the immensely facilitated means of communication, draws all, even the most barbarian, nations into civilization. The cheap prices of its commodities are the heavy artillery with which it batters down all Chinese walls, with which it forces the barbarians' intensely obstinate hatred of foreigners to capitulate. It compels all nations, on pain of extinction, to adopt the bourgeois mode of production; it compels them to introduce what it calls civilization into their midst, i.e., to become bourgeois themselves. In a word, it creates a world after its own image.

The bourgeoisie has subjected the country to the rule of the towns. It has created enormous cities, has greatly increased the urban population as compared with the rural, and has thus rescued a considerable part of the population from the idiocy of rural life. Just as it has made the country dependent on the towns, so it has made barbarian and semi-barbarian countries dependent on the civilised ones, nations of peasants on nations of bourgeois, the East on the West.

The bourgeoisie keeps more and more doing away with the scattered state of the population, of the means of production, and of property. It has agglomerated population, centralized means of production, and has concentrated property in a few hands. The necessary consequence of this was political centralization. Independent, or but loosely connected provinces, with separate interests, laws, governments and systems of taxation, became lumped together in one nation, with one government, one code of laws, one national class-interest, one frontier and one customs-tariff.

The bourgeoisie, during its rule of scarce one hundred years, has created more massive and more colossal productive forces than have all preceding generations together. Subjection of Nature's forces to man, machinery, application of chemistry to industry and agriculture, steam-navigation, railways, electric telegraphs, clearing of whole continents for cultivation, canalization of rivers, whole populations conjured out of the ground—what earlier century had even a presentiment that such productive forces slumbered in the lap of social labor?

We see then: The means of production and of exchange on whose foundation the bourgeoisie built itself up, were generated in feudal society. At a certain stage in the development of these means of production and of exchange, the conditions under which feudal society produced and exchanged, the feudal organization of agriculture and manufacturing industry, in one word, the feudal relations of property became no longer compatible with the already developed productive forces; they became so many fetters. They had to burst asunder; they were burst asunder.

Into their places stepped free competition, accompanied by a social and political constitution adapted to it, and by the economical and political sway of the bourgeois class.

A similar movement is going on before our own eyes. Modern bourgeois society with its relations of production, of exchange and of property, a society that has conjured up such gigantic means of production and of exchange, is like the sorcerer, who is no longer able to control the powers of the nether world whom he has called up by his spells. For many a decade past the history of industry and commerce is but the history of the revolt of modern productive forces against modern conditions of production, against the property relations that are the conditions for the existence of the bourgeoisie and of its rule. It is enough to mention the commercial crises that by their periodical return put on its trial, each time more threateningly, the existence of the entire bourgeois society. In these crises a great part not only of the existing products, but also of the previously created productive forces, are periodically destroyed. In these crises there breaks out an epidemic that, in all earlier epochs, would have seemed an absurdity—the epidemic of overproduction. Society suddenly finds itself put back into a state of momentary barbarism; it appears as if a famine, a universal war of devastation had cut off the supply of every means of subsistence; industry and commerce seem to be destroyed; and why? Because there is too much civilization, too much means of subsistence, too much industry, too much commerce. The productive forces at the disposal of society no longer tend to further the development of the conditions of bourgeois property; on the contrary, they have become too powerful for these conditions, by which they are fettered, and so soon as they overcome these fetters, they bring disorder into the whole of bourgeois society, endanger the existence of bourgeois property. The conditions of bourgeois society are too narrow to comprise the wealth created by them. And how does the bourgeoisie get over these crises? On the one hand by enforced destruction of a mass of productive forces; on the other, by the conquest of new markets, and by the more thorough exploitation of the old ones. That is to say, by paving the way for more extensive and more destructive crises, and by diminishing the means whereby crises are prevented.

The weapons with which the bourgeoisie felled feudalism to the ground are now turned against the bourgeoisie itself.

But not only has the bourgeoisie forged the weapons that bring death to itself; it has also called into existence the men who are to wield those weapons—the modern working class—the proletarians.

In proportion as the bourgeoisie, i.e., capital, is developed, in the same proportion is the proletariat, the modern working class, developed, a class of laborers, who live only so long as they find work, and who find work only so long as their labor increases capital. These laborers, who must sell themselves piecemeal, are a commodity, like every other article of commerce, and are consequently exposed to all the vicissitudes of competition, to all the fluctuations of the market.

Owing to the extensive use of machinery and to division of labor, the work of the proletarians has lost all individual character, and, consequently, all charm for the workman. He becomes an appendage of the machine, and it is only the most simple, most monotonous, and most easily acquired knack that is required of him. Hence, the cost of production of a workman is restricted, almost entirely, to the means of subsistence that he requires for his maintenance, and for the propagation of his race. But the price of a commodity, and also of labor, is equal to its cost of production. In proportion, therefore, as the repulsiveness of the work increases, the wage decreases. Nay more, in proportion as the use of machinery and division of labor increases, in the same proportion the burden of toil also increases, whether by prolongation of the working hours, by increase of the work enacted in a given time, or by increased speed of the machinery, etc.

Modern industry has converted the little workshop of the patriarchal master in the great factory of the industrial capitalist. Masses of laborers, crowded into the factory, are organized like soldiers. As privates of the industrial army they are placed under the command of a perfect hierarchy of officers and sergeants. Not only are they the slaves of the bourgeois class, and of the bourgeois State, they are daily and hourly enslaved by the machine, by the over-looker, and, above all, by the individual bourgeois manufacturer himself. The more openly this despotism proclaims gain to be its end and aim, the more petty, the more hateful and the more embittering it is.

The less the skill and exertion or strength implied in manual labor, in other words, the more modern industry becomes developed, the more is the labor of men superseded by that of women. Differences of age and sex have no longer any distinctive social validity for the working class. All are instruments of labor, more or less expensive to use, according to their age and sex.

No sooner is the exploitation of the laborer by the manufacturer, so far, at an end, that he receives his wages in cash, than he is set upon by the other portions of the bourgeoisie, the landlord, the shopkeeper, the pawnbroker, etc.

The lower strata of the middle class—the small tradespeople, shopkeepers, and retired tradesmen generally, the handicraftsmen and peasants—all these sink gradually into the proletariat, partly because their diminutive capital does not suffice for the scale on which Modern Industry is carried on, and is swamped in the competition with the large capitalists, partly because their specialised skill is rendered worthless by new methods of production. Thus the proletariat is recruited from all classes of the population.

The proletariat goes through various stages of development. With its birth begins its struggle with the bourgeoisie. At first the contest is carried on by individual laborers, then by the workpeople of a factory, then by the operatives of one trade, in one locality, against the individual bourgeois who directly exploits them. They direct their attacks not against the bourgeois conditions of production, but against the instruments

of production themselves; they destroy imported wares that compete with their labor, they smash to pieces machinery, they set factories ablaze, they seek to restore by force the vanished status of the workman of the Middle Ages.

At this stage the laborers still form an incoherent mass scattered over the whole country, and broken up by their mutual competition. If anywhere they unite to form more compact bodies, this is not yet the consequence of their own active union, but of the union of the bourgeoisie, which class, in order to attain its own political ends, is compelled to set the whole proletariat in motion, and is moreover yet, for a time, able to do so. At this stage, therefore, the proletarians do not fight their enemies, but the enemies of their enemies, the remnants of absolute monarchy, the landowners, the non-industrial bourgeois, the petty bourgeoisie. Thus the whole historical movement is concentrated in the hands of the bourgeoisie; every victory so obtained is a victory for the bourgeoisie.

But with the development of industry the proletariat not only increases in number, it becomes concentrated in greater masses, its strength grows, and it feels that strength more. The various interests and conditions of life within the ranks of the proletariat are more and more equalized, in proportion as machinery obliterates all distinctions of labor, and nearly everywhere reduces wages to the same low level. The growing competition among the bourgeois, and the resulting commercial crises, make the wages of the workers ever more fluctuating. The unceasing improvement of machinery, ever more rapidly developing, makes their livelihood more and more precarious, the collisions between individual workmen and individual bourgeois take more and more the character of collisions between two classes. Thereupon the workers begin to form combinations (Trades' Unions) against the bourgeois; they club together in order to keep up the rate of wages; they found permanent associations in order to make provision beforehand for these occasional revolts. Here and there the contest breaks out into riots.

Now and then the workers are victorious, but only for a time. The real fruit of their battles lies, not in the immediate result, but in the ever expanding union of the workers. This union is helped on by the improved means of communication that are created by modern industry, and that place the workers of different localities in contact with one another. It was just this contact that was needed to centralise the numerous local struggles, all of the same character, into one national struggle between classes. But every class struggle is a political struggle. And that union, to attain which the burghers of the Middle Ages, with their miserable highways, required centuries, the modern proletarians, thanks to railways, achieve in a few years.

This organization of the proletarians into a class, and consequently into a political party, is continually being upset again by the competition between the workers themselves. But it ever rises up again, stronger, firmer, mightier. It compels legislative recognition of particular interests of the workers, by taking advantage of the divisions among the bourgeoisie itself. Thus the ten-hours'-bill in England was carried.

Altogether collisions between the classes of the old society further, in many ways, the course of development of the proletariat. The bourgeoisie finds itself involved in a constant battle. At first with the aristocracy; later on, with those portions of the bourgeoisie itself, whose interests have become antagonistic to the progress of industry; at all times, with the bourgeoisie of foreign countries. In all these battles it sees itself compelled to appeal to the proletariat, to ask for its help, and thus, to drag it into the political arena. The bourgeoisie itself, therefore, supplies the proletariat with its own elements of political and general education, in other words, it furnishes the proletariat with weapons for fighting the bourgeoisie.

Further, as we have already seen, entire sections of the ruling classes are, by the advance of industry, precipitated into the proletariat, or are at least threatened in their conditions of existence. These also supply the proletariat with fresh elements of enlightenment and progress.

Finally, in times when the class-struggle nears the decisive hour, the process of dissolution going on within the ruling class, in fact within the whole range of old society, assumes such a violent, glaring character, that a small section of the ruling class cuts itself adrift, and joins the revolutionary class, the class that holds the future in its hands. Just as, therefore, at an earlier period, a section of the nobility went over to the bourgeoisie, so now a portion of the bourgeoisie goes over to the proletariat, and in particular, a portion of the bourgeois ideologists, who have raised themselves to the level of comprehending theoretically the historical movements as a whole.

Of all the classes that stand face to face with the bourgeoisie today, the proletariat alone is a really revolutionary class. The other classes decay and finally disappear in the face of modern industry; the proletariat is its special and essential product.

The lower-middle class, the small manufacturer, the shopkeeper, the artisan, the peasant, all these fight against the bourgeoisie, to save from extinction their existence as fractions of the middle class. They are therefore not revolutionary, but conservative. Nay more, they are reactionary, for they try to roll back the wheel of history. If by chance they are revolutionary, they are so, only in view of their impending transfer into the proletariat, they thus defend not their present, but their future interests, they desert their own standpoint to place themselves at that of the proletariat.

The "dangerous class," the social scum, that passively rotting mass thrown off by the lowest layers of old society, may, here and there, be swept into the movement by a proletarian revolution; its conditions of life, however, prepare it far more for the part of a bribed tool of reactionary intrigue.

In the conditions of the proletariat, those of old society at large are already virtually swamped. The proletarian is without property; his relation to his wife and children has no longer anything in common with the bourgeois family-relations; modern industrial labor, modern subjection to capital, the same in England as in France, in America as in Germany, has stripped him of every trace of national character. Law, morality, religion, are to him so many bourgeois prejudices, behind which lurk in ambush just as many bourgeois interests.

All the preceding classes that got the upper hand, sought to fortify their already acquired status by subjecting society at large to their conditions of appropriation. The proletarians cannot become masters of the productive forces of society, except by abolishing their own previous mode of appropriation, and thereby also every other previous mode of appropriation. They have nothing of their own to secure and to fortify; their mission is to destroy all previous securities for, and insurances of, individual property.

All previous historical movements were movements of minorities, or in the interest of minorities. The proletarian movement is the self-conscious, independent movement of the immense majority, in the interest of the immense majority. The proletariat, the lowest stratum of our present society, cannot stir, cannot raise itself up, without the whole superincumbent strata of official society being sprung into the air.

Though not in substance, yet in form, the struggle of the proletariat with the bourgeoisie is at first a national struggle. The proletariat of each country must, of course, first of all settle matters with its own bourgeoisie.

In depicting the most general phases of the development of the proletariat, we traced the more or less veiled civil war, raging within existing society, up to the point where that war breaks out into open revolution, and where the violent overthrow of the bourgeoisie, lays the foundation for the sway of the proletariat.

Hitherto, every form of society has been based, as we have already seen, on the antagonism of oppressing and oppressed classes. But in order to oppress a class, certain conditions must be assured to it under which it can, at least, continue its slavish existence. The serf, in the period of serfdom, raised himself to membership in the commune, just as the petty bourgeois, under the yoke of feudal absolutism, managed to develop into a bourgeois. The modern laborer, on the contrary, instead of rising with the progress of industry, sinks deeper and deeper below the conditions of existence of his own class. He becomes a pauper, and pauperism develops more rapidly than population and wealth. And here it becomes evident, that the bourgeoisie is unfit any longer to be the ruling class in society, and to impose its conditions of existence upon society as an overriding law. It is unfit to rule, because it is incompetant to assure an existence to its slave within his slavery, because it cannot help letting him sink into such a state, that it has to feed him, instead of being fed by him. Society can no longer live under this bourgeoisie, in other words, its existence is no longer compatible with society.

The essential condition for the existence, and for the sway of the bourgeois class, is the formation and augmentation of capital; the condition for capital is wage-labor. Wage-labor rests exclusively on competition between the laborers. The advance of industry, whose involuntary promoter is the bourgeoisie, replaces the isolation of the laborers, due to competition, by their involuntary combination, due to association. The development of Modern Industry, therefore, cuts from under its feet the very foundation on which the bourgeoisie produces and appropriates products. What the bourgeoisie therefore produces, above all, are its own grave-diggers. Its fall and the victory of the proletariat are equally inevitable.

NOTES

1. By *bourgeoisie* is meant the class of modern capitalists, owners of the means of social production and employers of wage-labor. By proletariat, the class of modern wage-laborers who, having no means of production of their own, are reduced to selling their labor-power in order to live.

2. That is, all written history. In 1847, the prehistory of society, the social organization existing previous to recorded history, was all but unknown. Since then, Haxthausen discovered common ownership of land in Russia. Maurer proved it to be the social foundation from which all Teutonic races started in history,

and by and by village communities were found to be, or to have been, the primitive form of society everywhere from India to Ireland. The inner organization of this primitive Communistic society was laid bare, in its typical form, by Morgan's crowning discovery of the true nature of the gens and its relationship to the tribe. With the dissolution of these primaeval communities society begins to be differentiated into separate and finally antagonistic classes. I have attempted to retrace this process of dissolution in: "Der Ursprung der Familie des, Privatelgenthums und des Staats," 2d ed., Stuttgart 1886.

3. Guild-master, that is a full member of a guild, a master within, not head of, a guild.

4. "Commune" was the name taken, in France, by the nascent towns even before they had conquered from their feudal lords and masters, local self-government and political rights as "the Third Estate". Generally speaking, for the economical development of the bourgeoisie, England is here taken as the typical country, for its political development, France.

DISCUSSION QUESTIONS

1. Marx and Engels see class conflict as inherent in capitalist societies. What changes in class relations do they envision over time? Do you see evidence of this in the contemporary world?

2. What do Marx and Engels mean when they say that the bourgeoisie class turns all social relations into money relations? What evidence do you see of this? Marx and Engels argue that an ever-expanding market (and, by implication, increased consumption) is the driving force of capitalism.

3. Marx and Engels were writing over 150 years ago. What evidence do you see now of increasing consumerism and how this shapes the values of people in the United States? What effect does it have on the class system?

4. According to Marx and Engels, the bourgeoisie is driven to create ever-new markets, and it draws nations of the world into this capitalist market. Were Marx and Engels writing in the 21st century, what kinds of similar phenomena might they include in a newly written essay?

INTERNET RESOURCES

Suggested Web URLS
for Further Study

http://csf.colorado.edu/psn/marx/index.html
This site contains a library, a biographical archive, a photo gallery, other Marxist writers and a search engine that searches the entire Marx/Engels Internet Library.

http://csf.colorado.edu/psn/seminars/manifesto.html
This Web site contains a paper entitled *The Making of the Manifesto* presented for the virtual seminar of the Progressive Sociologists' Network.

InfoTrac College Edition

You can find further relevant readings on the World Wide Web at *http://sociology.wadsworth.com*

Virtual Society

For further information on this subject including links to relevant Web sites, go to the Wadsworth Sociology homepage at *http://sociology.wadsworth.com*

4

How to Lie with Statistics

DARRELL HUFF

'll face up to the serious purpose . . . beneath the surface . . . : explaining how to look a phony statistic in the eye and face it down; and no less important, how to recognize sound and usable data in that wilderness of fraud. . . .

Not all the statistical information that you may come upon can be tested with the sureness of chemical analysis or of what goes on in an assayer's laboratory. But you can prod the stuff with four simple questions, and by finding the answers avoid learning a remarkable lot that isn't so.

WHO SAYS SO?

About the first thing to look for is bias—the laboratory with something to prove for the sake of a theory, a reputation, or a fee; the newspaper whose aim is a good story; labor or management with a wage level at stake.

Look for conscious bias. The method may be direct misstatement or it may be ambiguous statement that serves as well and cannot be convicted. It may be selection of favorable data and suppression of unfavorable. Units of measurement may be shifted, as with the practice of using one year for one comparison and sliding over to a more favorable year for another. An improper measure may be used: a mean where a median would be more informative (perhaps all too informative), with the trickery covered by the unqualified word "average."

Look sharply for unconscious bias. It is often more dangerous. In the charts and predictions of many statisticians and economists in 1928 it operated to produce remarkable things. The cracks in the economic structure were joyously overlooked, and all sorts of evidence was adduced and statistically supported to show that we had no more than entered the stream of prosperity.

It may take at least a second look to find out who-says-so. The who may be hidden by what Stephen Potter, the *Lifemanship* man, would probably call the "O.K. name." Anything smacking of the medical profession is an O.K. name. Scientific laboratories have O.K. names. So do colleges, especially universities, more especially ones eminent in technical work. . . . Please note that while the data came from Cornell, the conclusions were entirely the writer's own. But the O.K. name helps you carry away a misimpression of "Cornell University says . . ."

When an O.K. name is cited, make sure that the authority stands behind the information, not merely somewhere alongside it. . . .

HOW DOES HE KNOW?

It turns out that the *Journal* had begun by sending its questionnaires to 1,200 large companies. Only fourteen per cent had replied. Eighty-six per cent had not cared to say anything in public on whether they were hoarding or price gouging.

The *Journal* had put a remarkably good face on things, but the fact remains that there was little to brag about. It came down to this: Of 1,200 companies polled, nine per cent said they had not raised prices, five per cent said they had, and eighty-six per cent wouldn't say. Those that had replied constituted a sample in which bias might be suspected.

Watch out for evidence of a biased sample, one that has been selected improperly or—as with this one—has selected itself. Ask the question we dealt with in an early chapter: Is the sample large enough to permit any reliable conclusion?

Similarly with a reported correlation: Is it big enough to mean anything? Are there enough cases to add up to any significance? You cannot, as a casual reader, apply tests of significance or come to exact conclusions as to the adequacy of a sample. On a good many of the things you see reported, however, you will be able to tell at a glance—a good long glance, perhaps—that there just weren't enough cases to convince any reasoning person of anything.

You won't always be told how many cases. The absence of such a figure, particularly when the source is an interested one, is enough to throw suspicion on the whole thing. Similarly a correlation given without a measure of reliability (probable error, standard error) is not to be taken very seriously.

Watch out for an average, variety unspecified, in any matter where mean and median might be expected to differ substantially.

Many figures lose meaning because a comparison is missing. An article in *Look* magazine says, in connection with Mongolism, that "one study shows that in 2,800 cases, over half of the mothers were 35 or over." Getting any meaning from this depends upon your knowing something about the ages at which women in general produce babies. Few of us know things like that. . . .

Sometimes it is percentages that are given and raw figures that are missing, and this can be deceptive too. Long ago, when Johns Hopkins University had just begun to admit women students, someone not particularly enamored of coeducation reported a real shocker: Thirty-three and one-third per cent of the women at Hopkins had married faculty members! The raw figures gave a clearer picture. There were three women enrolled at the time, and one of them had married a faculty man. . . .

Sometimes what is missing is the factor that caused a change to occur. This omission leaves the implication that some other, more desired, factor is responsible. Figures published one year attempted to show that business was on the upgrade by pointing out that April retail sales were greater than in the year before. What was missing was the fact that Easter had come in March in the earlier year and in April in the later year.

A report of a great increase in the deaths from cancer in the last quarter-century is misleading unless you know how much of it is a product of such extraneous factors as these: Cancer is often listed now where "causes unknown" was formerly used; autopsies are more frequent, giving surer diagnoses; reporting and compiling of medical statistics are more complete; and people more frequently reach the most susceptible ages now. And if you are looking at total deaths rather than the death rate, don't neglect the fact that there are more people now than there used to be.

DID SOMEBODY
CHANGE THE SUBJECT?

When assaying a statistic, watch out for a switch somewhere between the raw figure and the conclusion. One thing is all too often reported as another.

As just indicated, more reported cases of a disease are not always the same thing as more cases of the disease. A straw-vote victory for a candidate is not always negotiable at the polls. An expressed preference by a "cross-section" of a magazine's readers for articles on world affairs is no final proof that they would read the articles if they were published.

Encephalitis cases reported in the central valley of California in 1952 were triple the figure for the worst previous year. Many alarmed residents shipped their children away. But when the reckoning was in, there had been no great increase in deaths from sleeping sickness. What had happened was that state and federal health people had come in in great numbers to tackle a long-time problem; as a result of their efforts a great many low-grade cases were recorded that in other years would have been overlooked, possibly not even recognized. . . .

"The British male over 5 years of age soaks himself in a hot tub on an average of 1.7 times a week in the winter and 2.1 times in the summer," says a newspaper story. "British women average 1.5 baths a week in the winter and 2.0 in the summer." The source is a Ministry of Works hot-water survey of "6,000 representative British homes." The sample was representative, it says, and seems quite adequate in size to justify the conclusion in the San Francisco *Chronicle's* amusing headline: BRITISH HE'S BATHE MORE THAN SHE'S.

The figures would be more informative if there were some indication of whether they are means or medians. However, the major weakness is that the subject has been changed. What the Ministry really found out is how often these people said they bathed, not how often they did so. When a subject is as intimate as this one is, with the British bath-taking tradition involved, saying and doing may not be the same thing at all. British he's may or may not bathe oftener than she's; all that can safely be concluded is that they say they do.

Here are some more varieties of change-of-subject to watch out for.

A back-to-the-farm movement was discerned when a census showed half a million more farms in 1935 than five years earlier. But the two counts were not talking about the same thing. The definition of farm used by the Bureau of the Census had been changed; it took in at least 300,000 farms that would not have been so listed under the 1930 definition. . . .

The "population" of a large area in China was 28 million. Five years later it was 105 million. Very little of that increase was real; the great difference could be explained only by taking into account the purposes of the two enumerations and the way people would be inclined to feel about being counted in each instance. The first census was for tax and military purposes, the second for famine relief. . . .

The *post hoc* variety of pretentious nonsense is another way of changing the subject without seeming to. The change of something *with* something else is presented as *because of*. The magazine *Electrical World* once offered a composite chart in an editorial on "What Electricity Means to America." You could see from it that as "electrical horsepower in factories" climbed, so did "average wages per hour." At the same time "average hours per week" dropped. All these things are long-time trends, of course, and there is no evidence at all that any one of them has produced any other.

And then there are the firsters. Almost anybody can claim to be first in *something* if he is not too particular what it is. At the end of 1952 two New York newspapers were each insisting on first rank in grocery advertising. Both were right too, in a way. The *World-Telegram* went on to explain that it was first in full-run advertising, the kind that appears in all copies, which is the only kind it runs. The *Journal-American* insisted that total linage was what counted and that it was first in that. This is the kind of reaching for a superlative that leads the weather reporter on the radio to label a quite normal day "the hottest June second since 1949."

Change-of-subject makes it difficult to compare cost when you contemplate borrowing money either directly or in the form of installment buying. Six per cent sounds like six per cent—but it may not be at all.

If you borrow $100 from a bank at six per cent interest and pay it back in equal monthly installments for a year, the price you pay for the use of the money is about $3. But another six per cent loan, on the basis sometimes called $6 on the $100, will cost you twice as much. That's the way most automobile loans are figured. It is very tricky.

The point is that you don't have the $100 for a year. By the end of six months you have paid back half of it. If you are charged at $6 on the $100, or six percent of the amount, you really pay interest at nearly twelve per cent. . . .

Sometimes the semantic approach will be used to change the subject. Here is an item from *Business Week* magazine.

Accountants have decided that "surplus" is a nasty word. They propose eliminating it from corporate balance sheets. The Committee on Accounting Procedure of the American Institute of Accountants says: . . . Use such descriptive terms as "retained earnings" or "appreciation of fixed assets."

This one is from a newspaper story reporting Standard Oil's record-breaking revenue and net profit of a million dollars a day.

Possibly the directors may be thinking some time of splitting the stock for there may be an advantage . . . if the profits per share do not look so large. . . .

DOES IT MAKE SENSE?

"Does it make sense?" will often cut a statistic down to size when the whole rigmarole is based on an unproved assumption. You may be familiar with the Rudolf Flesch readability formula. It purports to measure how easy a piece of prose is to read, by such simple and objective items as length of words and sentences. Like all devices for reducing the imponderable to a number and substituting arithmetic for judgment, it is an appealing idea. At least it has appealed to people who employ writers, such as newspaper publishers, even if not to many writers themselves. The assumption in the formula is that such things as word length determine readability. This, to be ornery about it, remains to be proved.

A man named Robert A. Dufour put the Flesch formula to trial on some literature that he found handy. It showed "The Legend of Sleepy Hollow" to be half again as hard to read as Plato's *Republic*. The Sinclair Lewis novel *Cass Timberlane* was rated more difficult than an essay by Jacques Maritain, "The Spiritual Value of Art." A likely story.

Many a statistic is false on its face. It gets by only because the magic of numbers brings about a suspension of common sense. Leonard Engel, in a *Harper's* article, has listed a few of the medical variety.

An example is the calculation of a well-known urologist that there are eight million cases of cancer of the prostate gland in the United States—which would be enough to provide 1.1 carcinomatous prostate glands for every male in the susceptible age group! Another is a prominent neurologist's estimate that one American in twelve suffers from migraine; since migraine is responsible for a third of chronic headache cases, this would mean that a quarter of us must suffer from disabling headaches. Still another is the figure of 250,000 often given for the number of multiple sclerosis cases; death data indicate that there can be, happily, no more than thirty to forty thousand cases of this paralytic disease in the country.

Hearings on amendments to the Social Security Act have been haunted by various forms of a statement that makes sense only when not looked at closely. It is an argument that goes like this: Since life expectancy is only about sixty-three years, it is a sham and a fraud to set up a social-security plan with a retirement age of sixty-five, because virtually everybody dies before that.

You can rebut that one by looking around at people you know. The basic fallacy, however, is that the figure refers to expectancy at birth, and so about half the babies born can expect to live longer than that. The figure, incidentally, is from the latest official complete life table and is correct for the 1939–1941 period. An up-to-date estimate corrects it to sixty-five-plus. Maybe that will produce a new and equally silly argument to the effect that practically everybody now lives to be sixty-five. . . .

You are entitled to look with the same suspicion on the report, some years ago, by the American Petroleum Industries Committee that the average yearly tax bill for automobiles is $51.13.

Extrapolations are useful, particularly in that form of soothsaying called forecasting trends. But in looking at the figures on the charts made from them, it is necessary to remember one thing constantly: The trend-to-now may be a fact, but the future trend represents no more than an educated guess. Implicit in it is "everything else being equal" and "present trends continuing." And somehow everything else refuses to remain equal, else life would be dull indeed. . . .

. . . in 1874, Mark Twain summed up the nonsense side of extrapolation in *Life on the Mississippi:*

In the space of one hundred and seventy-six years the Lower Mississippi has shortened itself two hundred and forty-two miles. That is an average of a trifle over one mile and a third per year. Therefore, any calm person, who is not blind or idiotic, can see that in the Old Oölitic Silurian Period, just a million years ago next November, the Lower Mississippi River was upward of one million three hundred thousand miles long, and stuck out over the Gulf of Mexico like a fishing-rod. And by the same token any person can see that seven hundred and forty-two years from now that Lower Mississippi will be only a mile and three-quarters long, and Cairo and New Orleans will have joined their streets together, and be plodding comfortably along under a single mayor and a mutual board of aldermen. There is something fascinating about science. One gets such wholesale returns of conjecture out of such a trifling investment of fact.

DISCUSSION QUESTIONS

1. Peruse the daily paper (or magazine) and find a report about some new study that makes a statistical or numerical claim. What claim does the headline make? Now ask yourself if these claims are reliable after you answer the four questions Huff poses.

2. Picking from some popular media form, take two research claims based on statistics. Elect one that you find credible and one that seems suspicious in its claims. What makes one credible and the other not? How do Huff's arguments help you decide which statistical "facts" are reliable?

3. We live in an increasingly information-based society. Why do we put so much faith in numerical arguments in such a society? What has led to the development of such a society? How do you see this evolving in the future?

4. Knowledge is power, so the saying goes. In an information-based society those without numerical literacy (i.e., the ability to understand and interpret numerical claims) risk being powerless. What groups are most vulnerable to this? What are the sociological factors that create this vulnerability?

INTERNET RESOURCES

Suggested Web URLS
for Further Study

http://www.hartford-hwp.com/archives/index.html
World History Archives. Online resources and historical facts from around the world.

http://www.socsciresearch.com/
Research Resources for the Social Sciences.

http://www.odci.gov/cia/publications/factbook/
The World Factbook 2000. Hosted by the CIA. Contains basic facts from around the world.

InfoTrac College Edition

You can find further relevant readings on the World Wide Web at
http://sociology.wadsworth.com

Virtual Society

For further information on this subject including links to relevant Web sites, go to the Wadsworth Sociology homepage at
http://sociology.wadsworth.com

5

Manifest and Latent Functions

ROBERT K. MERTON

Armed with the concept of latent function, the sociologist extends his inquiry in those very directions which promise most for the theoretic development of the discipline. He examines the familiar (or planned) social practice to ascertain the latent, and hence generally unrecognized, functions (as well, of course, as the manifest functions). He considers, for example, the consequences of the new wage plan for, say, the trade union in which the workers are organized or the consequences of a propaganda program, not only for increasing its avowed purpose of stirring up patriotic fervor, but also for making large numbers of people reluctant to speak their minds when they differ with official policies, etc. In short, it is suggested that the *distinctive* intellectual contributions of the sociologist are found primarily in the study of unintended consequences (among which are latent functions) of social practices, as well as in the study of anticipated consequences (among which are manifest functions).

The manifest purpose of buying consumption goods is, of course, the satisfaction of the needs for which these goods are explicitly designed.

Thus, automobiles are obviously intended to provide a certain kind of transportation; candles, to provide light; choice articles of food to provide sustenance; rare art products to provide aesthetic pleasure. Since these products *do* have these uses, it was largely assumed that these encompass the range of socially significant functions. Veblen indeed suggests that this was ordinarily the prevailing view (in the pre-Veblenian era, of course): "The end of acquisition and accumulation is conventionally held to be the consumption of the goods accumulated. . . . This is at least felt to be the economically legitimate end of acquisition, *which alone it is incumbent on the theory to take account of.*"[1]

However, says Veblen in effect, as sociologists we must go on to consider the latent functions of acquisition, accumulation, and consumption, and these latent functions are remote indeed from the manifest functions. "But, it is only when taken in a sense far removed from its naive meaning [i.e., manifest function] that the consumption of goods can be said to afford the incentive from which accumulation invariably proceeds." And among these latent functions, which help explain the persistence and the social

Credit: "Manifest and Latent Functions," by Robert K. Merton from *Social Theory and Social Structure* by Robert Merton ©1967, 1968 by Robert K. Merton. Reprinted by permission of The Free Press.

location of the pattern of conspicuous consumption, it [the fact that] . . . it results in *a heightening* or *reaffirmation of social status.*

The Veblenian paradox is that people buy expensive goods not so much because they are superior but because they are expensive. For it is the latent equation ("costliness = mark of higher social status") which he singles out in his functional analysis, rather than the manifest equation ("costliness = excellence of the goods"). Not that

he denies manifest functions *any* place in buttressing the pattern of conspicuous consumption. These, too, are operative. . . . *It is only that these direct, manifest functions do not fully account for the prevailing patterns of consumption. Otherwise put, if the latent functions of status-enhancement or status-reaffirmation were removed from the patterns of conspicuous consumption, these patterns would undergo severe changes of a sort which the "conventional" economist could not foresee.*

NOTE

1.   Thorstein Veblen, *Theory of the Leisure Class* (1899) (New York: Vanguard Press, 1928), p. 25.

DISCUSSION QUESTIONS

1.   What are the manifest functions of a child's allowance? What are the latent functions?

2.   How would you use Merton's analysis of manifest and latent functions to develop a sociological analysis of the fashion industry?

3.   Merton's work is based in a theoretical framework of functionalism. What are the key points in functionalist theory, and how does his discussion of manifest and latent functions stem from this sociological theory?

INTERNET RESOURCES

Suggested Web URLs
for Further Study

http://xroads.virginia.edu/~HYPER/VEBLEN /veblenhp.html
Veblen, Thorstein: *The Theory of the Leisure Class.* On-line table of contents and full text of this classic piece.

http://www.pscw.uva.nl/sociosite/topics/sociologists.ht ml#MERTON
This Web site contains links to works by Robert K. Merton, as well as links to papers commenting on Merton's works.

InfoTrac College Edition

You can find further relevant readings on the World Wide Web at
http://sociology.wadsworth.com

Virtual Society

For further information on this subject including links to relevant Web sites, go to the Wadsworth Sociology homepage at
http://sociology.wadsworth.com

6

Body Ritual among the Nacirema

HORACE MINER

The anthropologist has become so familiar with the diversity of ways in which different peoples behave in similar situations that he is not apt to be surprised by even the most exotic customs. In fact, if all of the logically possible combinations of behavior have not been found somewhere in the world, he is apt to suspect that they must be present in some yet undescribed tribe. This point has, in fact, been expressed with respect to clan organization by Murdock (1949: 71). In this light, the magical beliefs and practices of the Nacirema present such unusual aspects that it seems desirable to describe them as an example of the extremes to which human behavior can go.

Professor Linton first brought the ritual of the Nacirema to the attention of anthropologists twenty years ago (1936: 326), but the culture of this people is still very poorly understood. They are a North American group living in the territory between the Canadian Cree, the Yaqui and Tarahumare of Mexico, and the Carib and Arawak of the Antilles. Little is known of their origin, although tradition states that they came from the east. According to Nacirema mythology, their nation was originated by a culture hero, Notgnihsaw, who is otherwise known for two great feats of strength—the throwing of a piece of wampum across the river Pa-To-Mac and the chopping down of a cherry tree in which the Spirit of Truth resided.

Nacirema culture is characterized by a highly developed market economy which has evolved in a rich natural habitat. While much of the people's time is devoted to economic pursuits, a large part of the fruits of these labors and a considerable portion of the day are spent in ritual activity. The focus of this activity is the human body, the appearance and health of which loom as a dominant concern in the ethos of the people. While such concern is certainly not unusual, its ceremonial aspects and associated philosophy are unique.

The fundamental belief underlying the whole system appears to be that the human body is ugly and that its natural tendency is to debility and disease. Incarcerated in such a body, man's only hope is to avert these characteristics through the use of the powerful influences of ritual and ceremony. Every household has one or more shrines devoted to this purpose. The more powerful individuals in this society have several shrines in their houses, and, in fact, the opulence of a house is

Credit: "Body Ritual Among the Nacirema," by Horace Miner, reproduced by permission of the American Anthropological Association from *American Anthropologist* 58:3, June 1956. Not for further reproduction.

often referred to in terms of the number of such ritual centers it possesses. Most houses are of wattle and daub construction, but the shrine rooms of the more wealthy are walled with stone. Poorer families imitate the rich by applying pottery plaques to their shrine walls.

While each family has at least one such shrine, the rituals associated with it are not family ceremonies but are private and secret. The rites are normally only discussed with children, and then only during the period when they are being initiated into these mysteries. I was able, however, to establish sufficient rapport with the natives to examine these shrines and to have the rituals described to me.

The focal point of the shrine is a box or chest which is built into the wall. In this chest are kept the many charms and magical potions without which no native believes he could live. These preparations are secured from a variety of specialized practitioners. The most powerful of these are the medicine men, whose assistance must be rewarded with substantial gifts. However, the medicine men do not provide the curative potions for their clients, but decide what the ingredients should be and then write them down in an ancient and secret language. This writing is understood only by the medicine men and by the herbalists who, for another gift, provide the required charm.

The charm is not disposed of after it has served its purpose, but is placed in the charm-box of the household shrine. As these magical materials are specific for certain ills, and the real or imagined maladies of the people are many, the charm-box is usually full to overflowing. The magical packets are so numerous that people forget what their purposes were and fear to use them again. While the natives are very vague on this point, we can only assume that the idea in retaining all the old magical materials is that their presence in the charm-box, before which the body rituals are conducted, will in some way protect the worshipper.

Beneath the charm-box is a small font. Each day every member of the family, in succession, enters the shrine room, bows his head before the charm-box, mingles different sorts of holy water in the font, and proceeds with a brief rite of ablution. The holy waters are secured from the Water Temple of the community, where the priests conduct elaborate ceremonies to make the liquid ritually pure.

In the hierarchy of magical practitioners, and below the medicine men in prestige, are specialists whose designation is best translated "holy-mouth-men." The Nacirema have an almost pathological horror of and fascination with the mouth, the condition of which is believed to have a supernatural influence on all social relationships. Were it not for the rituals of the mouth, they believe that their teeth would fall out, their gums bleed, their jaws shrink, their friends desert them, and their lovers reject them. They also believe that a strong relationship exists between oral and moral characteristics. For example, there is a ritual ablution of the mouth for children which is supposed to improve their moral fiber.

The daily body ritual performed by everyone includes a mouth-rite. Despite the fact that these people are so punctilious about care of the mouth, this rite involves a practice which strikes the uninitiated stranger as revolting. It was reported to me that the ritual consists of inserting a small bundle of hog hairs into the mouth, along with certain magical powders, and then moving the bundle in a highly formalized series of gestures.

In addition to the private mouth-rite, the people seek out a holy-mouth-man once or twice a year. These practitioners have an impressive set of paraphernalia, consisting of a variety of augers, awls, probes, and prods. The use of these objects in the exorcism of the evils of the mouth involves almost unbelievable ritual torture of the client. The holy-mouth-man opens the client's mouth and, using the above-mentioned tools, enlarges any holes which decay may have created in the teeth. Magical materials are put into these holes. If there are no naturally occurring holes in the teeth, large sections of one or more teeth are gouged out so that the supernatural substance can be applied. In the client's view, the purpose of these ministrations is to arrest decay and to draw friends. The extremely sacred and traditional character of the rite is evident in the fact that the natives return to the holy-mouth-man year after year, despite the fact that their teeth continue to decay.

It is to be hoped that, when a thorough study of the Nacirema is made, there will be careful inquiry into the personality structure of these people. One has but to watch the gleam in the eye of a holy-mouth-man, as he jabs an awl into an exposed nerve, to suspect that a certain amount of sadism is involved. If this can be established, a very interesting pattern emerges, for most of the population shows definite masochistic tendencies. It was to these that Professor Linton referred in discussing a distinctive part of the daily body ritual which is performed only by men. This part of the rite involves scraping and lacerating the surface of the face with a sharp instrument. Special women's rites are performed only four times during each lunar month, but what they lack in frequency is made up in barbarity. As part of this ceremony, women bake their heads in small ovens for about an hour. The theoretically interesting point is that what seems to be a preponderantly masochistic people have developed sadistic specialists.

The medicine men have an imposing temple, or *latipso,* in every community of any size. The more elaborate ceremonies required to treat very sick patients can only be performed at this temple. These ceremonies involve not only the thaumaturge but a permanent group of vestal maidens who move sedately about the temple chambers in distinctive costume and headdress.

The *latipso* ceremonies are so harsh that it is phenomenal that a fair proportion of the really sick natives who enter the temple ever recover. Small children whose indoctrination is still incomplete have been known to resist attempts to take them to the temple because "that is where you go to die." Despite this fact, sick adults are not only willing but eager to undergo the protracted ritual purification, if they can afford to do so. No matter how ill the supplicant or how grave the emergency, the guardians of many temples will not admit a client if he cannot give a rich gift to the custodian. Even after one has gained admission and survived the ceremonies, the guardians will not permit the neophyte to leave until he makes still another gift.

The supplicant entering the temple is first stripped of all his or her clothes. In everyday life the Nacirema avoids exposure of his body and its natural functions. Bathing and excretory acts are performed only in the secrecy of the household shrine, where they are ritualized as part of the body-rites. Psychological shock results from the fact that body secrecy is suddenly lost upon entry into the *latipso.* A man, whose own wife has never seen him in an excretory act, suddenly finds himself naked and assisted by a vestal maiden while he performs his natural functions into a sacred vessel. This sort of ceremonial treatment is necessitated by the fact that the excreta are used by a diviner to ascertain the course and nature of the client's sickness. Female clients, on the other hand, find their naked bodies are subjected to the scrutiny, manipulation, and prodding of the medicine men.

Few supplicants in the temple are well enough to do anything but lie on their hard beds. The daily ceremonies, like the rites of the holy-mouth-men, involve discomfort and torture. With ritual precision, the vestals awaken their miserable charges each dawn and roll them about on their beds of pain while performing ablutions, in the formal movements of which the maidens are highly trained. At other times they insert magic wands in the supplicant's mouth or force him to eat substances which are supposed to be healing. From time to time the medicine men come to their clients and jab magically treated needles into their flesh. The fact that these temple ceremonies may not cure, and may kill, the neophyte, in no way decreases the people's faith in the medicine men.

There remains one other kind of practitioner, known as a "listener." This witch-doctor has the power to exorcise the devils that lodge in the heads of people who have been bewitched. The Nacirema believe that parents bewitch their own children. Mothers are particularly suspected of putting a curse on children while teaching them the secret body rituals. The counter-magic of the witch-doctor is unusual in its lack of ritual. The patient simply tells the "listener" all his troubles and fears, beginning with the earliest difficulties he can remember. The memory displayed by the Nacirema in these exorcism sessions is truly remarkable. It is not uncommon for the patient to bemoan the rejection he felt upon being weaned as a babe, and a few individuals even see their troubles going back to the traumatic effects of their own birth.

In conclusion, mention must be made of certain practices which have their base in native esthetics but which depend upon the pervasive aversion to the natural body and its functions. There are ritual fasts to make fat people thin and ceremonial feasts to make thin people fat. Still other rites are used to make women's breasts larger if they are small, and smaller if they are large. General dissatisfaction with breast shape is symbolized in the fact that the ideal form is virtually outside the range of human variation. A few women afflicted with almost inhuman hyper-mammary development are so idolized that they make a handsome living by simply going from village to village and permitting the natives to stare at them for a fee.

Reference has already been made to the fact that excretory functions are ritualized, routinized, and relegated to secrecy. Natural reproductive functions are similarly distorted. Intercourse is taboo as a topic and scheduled as an act. Efforts are made to avoid pregnancy by the use of magical materials or by limiting intercourse to certain phases of the moon. Conception is actually very infrequent. When pregnant, women dress so as to hide their condition. Parturition takes place in secret, without friends or relatives to assist, and the majority of women do not nurse their infants.

Our review of the ritual life of the Nacirema has certainly shown them to be a magic-ridden people. It is hard to understand how they have managed to exist so long under the burdens which they have imposed upon themselves. But even such exotic customs as these take on real meaning when they are viewed with the insight provided by Malinowski when he wrote (1948: 70):

> Looking from far and above, from our high places of safety in the developed civilization, it is easy to see all the crudity and irrelevance of magic. But without its power and guidance early man could not have mastered his practical difficulties as he has done, nor could man have advanced to the higher stages of civilization.

REFERENCES

Linton, R. 1936. *The study of man.* New York: Appleton-Century.

Malinowski, B. 1948. *Magic, science and religion.* Glencoe, IL: Free Press.

Murdock, G. P. 1949. *Social structure.* New York: Macmillan.

DISCUSSION QUESTIONS

1. Miner's satirical piece is a good example of what Peter Berger calls "debunking." He debunks the ordinary, everyday cultural practices of Americans. Identify a different cultural practice and write a similar essay debunking it as if you were looking at it as a complete outsider. What do you learn from doing so?

2. What other kinds of rituals are common in American society, other than the cleansing rituals about which Miner writes? What function do such rituals have in American culture—other than their alleged purpose?

3. Miner presents the practices of the Nacirema as if they were magic. How might you see the same practices if you looked at them as if they stemmed solely from a scientific and technological worldview? What if you looked at them as a form of sport or drama? How does your view of behavior influence what you see?

INTERNET RESOURCES

Suggested Web URLs
for Further Study

http://www.beadsland.com/nacirema/#anthro
A Web site with further related readings on the Nacirema.

InfoTrac College Edition

You can find further relevant readings on the World Wide Web at
http://sociology.wadsworth.com

Virtual Society

For further information on this subject including links to relevant Web sites, go to the Wadsworth Sociology homepage at
http://sociology.wadsworth.com

7

The Self

GEORGE HERBERT MEAD

In our statement of the development of intelligence we have already suggested that the language process is essential for the development of the self. The self has a character which is different from that of the physiological organism proper. The self is something which has a development; it is not initially there, at birth, but arises in the process of social experience and activity, that is, develops in the given individual as a result of his relations to that process as a whole and to other individuals within that process . . .

We can distinguish very definitely between the self and the body. The body can be there and can operate in a very intelligent fashion without there being a self involved in the experience. The self has the characteristic that it is an object to itself, and that characteristic distinguishes it from other objects and from the body. It is perfectly true that the eye can see the foot, but it does not see the body as a whole. We cannot see our backs; we can feel certain portions of them, if we are agile, but we cannot get an experience of our whole body. There are, of course, experiences which are somewhat vague and difficult of location, but the bodily experiences are for us organized about a self. The foot and hand belong to the self. We can see our feet, especially if we look at them from the wrong end of an opera glass, as strange things which we have difficulty in recognizing as our own. The parts of the body are quite distinguishable from the self. We can lose parts of the body without any serious invasion of the self. The mere ability to experience different parts of the body is not different from the experience of a table. The table presents a different feel from what the hand does when one hand feels another, but it is an experience of something with which we come definitely into contact. The body does not experience itself as a whole, in the sense in which the self in some way enters into the experience of the self.

It is the characteristic of the self as an object to itself that I want to bring out. This characteristic is represented in the word "self," which is a reflexive, and indicates that which can be both subject and object. This type of object is essentially different from other objects, and in the past it has been distinguished as conscious, a term which indicates an experience with, an experience of, one's self. It was assumed that consciousness in some way carried this capacity of being an object to itself. In giving a behavioristic statement of consciousness we have to look for some sort of experience in which the physical organism can become an object to itself.[1]

When one is running to get away from someone who is chasing him, he is entirely occupied in this action, and his experience may be swallowed up in the objects about him, so that he has, at the time being, no consciousness of self at all. We must be, of course, very completely occupied to have that take place, but we can, I think, recognize that sort of a possible experience in which the self does not enter. We can, perhaps, get some light on that situation through those experiences in which in very intense action there appear in the experience of the individual, back of this intense action, memories and anticipations. Tolstoi as an officer in the war gives an account of having pictures of his past experience in the midst of his most intense action. There are also the pictures that flash into a person's mind when he is drowning. In such instances there is a contrast between an experience that is absolutely wound up in outside activity in which the self as an object does not enter, and an activity of memory and imagination in which the self is the principal object. The self is then entirely distinguishable from an organism that is surrounded by things and acts with reference to things, including parts of its own body. These latter may be objects like other objects, but they are just objects out there in the field, and they do not involve a self that is an object to the organism. This is, I think, frequently overlooked. It is that fact which makes our anthropomorphic reconstructions of animal life so fallacious. How can an individual get outside himself (experientially) in such a way as to become an object to himself? This is the essential psychological problem of selfhood or of self-consciousness; and its solution is to be found by referring to the process of social conduct or activity in which the given person or individual is implicated. The apparatus of reason would not be complete unless it swept itself into its own analysis of the field of experience; or unless the individual brought himself into the same experiential field as that of the other individual selves in relation to whom he acts in any given social situation. Reason cannot become impersonal unless it takes an objective, nonaffective attitude toward itself; otherwise we have just consciousness, not *self*-consciousness. And it is necessary to rational conduct that the individual should thus take an objective, impersonal attitude toward himself, that he should become an object to himself. For the individual organism is obviously an essential and important fact or constituent element of the empirical situation in which it acts; and without taking objective account of itself as such, it cannot act intelligently, or rationally.

The individual experiences himself as such, not directly, but only indirectly, from the particular standpoints of other individual members of the same social group, or from the generalized standpoint of the social group as a whole to which he belongs. For he enters his own experience as a self or individual, not directly or immediately, not by becoming a subject to himself, but only insofar as he first becomes an object to himself just as other individuals are objects to him or in his experience; and he becomes an object to himself only by taking the attitudes of other individuals toward himself within a social environment or context of experience and behavior in which both he and they are involved.

The importance of what we term "communication" lies in the fact that it provides a form of behavior in which the organism or the individual may become an object to himself. It is that sort of communication which we have been discussing—not communication in the sense of the cluck of the hen to the chickens, or the bark of a wolf to the pack, or the lowing of a cow, but communication in the sense of significant symbols, communication which is directed not only to others but also to the individual himself. So far as that type of communication is a part of behavior it at least introduces a self. Of course, one may hear without listening; one may see things that he does not realize; do things that he is not really aware of. But it is where one does respond to that which he addresses to another and where that response of his own becomes a part of his conduct, where he not only hears himself but responds to himself, talks and replies to himself as truly as the other person replies to him, that we have behavior in which the individuals become objects to themselves. . . .

The self, as that which can be an object to itself, is essentially a social structure, and it arises in social experience. After a self has arisen, it in a certain sense provides for itself its social experiences, and so we can conceive of an absolutely solitary self. But it is impossible to conceive of a self arising outside of social experience. When it has arisen we can think of a person in solitary confinement for the rest of his life, but who still has himself as a companion, and is able to think and to converse with himself as he had communicated with others. That process to which I have just referred, of responding to one's self as another responds to it, taking part in one's own conversation with others, being aware of what one is saying and using that awareness of what one is saying to determine what one is going to say thereafter—that is a process with which we are all familiar. We are continually following up our own address to other persons by an understanding of what we are saying, and using that understanding in the direction of our continued speech. We are finding out what we are going to say, what we are going to do, by saying and doing, and in the process we are continually controlling the process itself. In the conversation of gestures what we say calls out a certain response in another and that in turn changes our own action, so that we shift from what we started to do because of the reply the other makes. The conversation of gestures is the beginning of communication. The individual comes to carry on a conversation of gestures with himself. He says something, and that calls out a certain reply in himself which makes him change what he was going to say. One starts to say something, we will presume an unpleasant something, but when he starts to say it he realizes it is cruel. The effect on himself of what he is saying checks him; there is here a conversation of gestures between the individual and himself. We mean by significant speech that the

action is one that affects the individual himself, and that the effect upon the individual himself is part of the intelligent carrying-out of the conversation with others. Now we, so to speak, amputate that social phase and dispense with it for the time being, so that one is talking to one's self as one would talk to another person.[2]

This process of abstraction cannot be carried on indefinitely. One inevitably seeks an audience, has to pour himself out to somebody. In reflective intelligence one thinks to act, and to act solely so that this action remains a part of a social process. Thinking becomes preparatory to social action. The very process of thinking is, of course, simply an inner conversation that goes on, but it is a conversation of gestures which in its completion implies the expression of that which one thinks to an audience. One separates the significance of what he is saying to others from the actual speech and gets it ready before saying it. He thinks it out, and perhaps writes it in the form of a book; but it is still a part of social intercourse in which one is addressing other persons and at the same time addressing one's self, and in which one controls the address to other persons by the response made to one's own gesture. That the person should be responding to himself is necessary to the self, and it is this sort of social conduct which provides behavior within which that self appears. I know of no other form of behavior that the linguistic in which the individual is an object to himself, and, so far as I can see, the individual is not a self in the reflexive sense unless he is an object to himself. It is this fact that gives a critical importance to communication, since this is a type of behavior in which the individual does so respond to himself.

We realize in everyday conduct and experience that an individual does not mean a great deal of what he is doing and saying. We frequently say that such an individual is not himself. We come away from an interview with a realization that we have left out important things, that there are parts of the self that did not get into what was said. What determines the amount of the self that gets into communication is the social experience itself. Of course, a good deal of the self does not need to get expression. We carry on a whole series of different relationships to different people. We are one thing to one man and another thing to another. There are parts of the self which exist only for the self in relationship to itself. We divide ourselves up in all sorts of different selves with reference to our acquaintances. We discuss politics with one and religion with another. There are all sorts of different selves answering to all sorts of different social reactions. It is the social process itself that is responsible for the appearance of the self; it is not there as a self apart from this type of experience.

A multiple personality is in a certain sense normal, as I have just pointed out. . . .

The unity and structure of the complete self reflects the unity and structure of the social process as a whole; and each of the elementary selves of which it is composed reflects the unity and structure of one of the various aspects of that process in which the individual is implicated. In other words, the various elementary selves which constitute, or are organized into, a complete self are the various aspects of the structure of that complete self answering to the various aspects of the structure of the social process as a whole; the structure of the complete self is thus a reflection of the complete social process. The organization and unification of a social group is identical with the organization and unification of any one of the selves arising within the social process in which that group is engaged, or which it is carrying on.[3]

. . . Another set of background factors in the genesis of the self is represented in the activities of play and the game. . . . We find in children . . . imaginary companions which a good many children produce in their own experience. They organize in this way the responses which they call out in other persons and call out also in themselves. Of course, this playing with an imaginary companion is only a peculiarly interesting phase of ordinary play. Play in this sense, especially the stage which precedes the organized games, is a play at something. A child plays at being a mother, at being a teacher, at being a policeman; that is, it is taking different roles, as we say. We have something that suggests this in what we call the play of animals: A cat will play with her kittens,

and dogs play with each other. Two dogs playing with each other will attack and defend, in a process which if carried through would amount to an actual fight. There is a combination of responses which checks the depth of the bite. But we do not have in such a situation the dogs taking a definite role in the sense that a child deliberately takes the role of another. This tendency on the part of children is what we are working with in the kindergarten where the roles which the children assume are made the basis for training. When a child does assume a role he has in himself the stimuli which call out that particular response or group of responses. He may, of course, run away when he is chased, as the dog does, or he may turn around and strike back just as the dog does in his play. But that is not the same as playing at something. Children get together to "play Indian." This means that the child has a certain set of stimuli that call out in itself the responses that they would call out in others, and which answer to an Indian. In the play period the child utilizes his own responses to these stimuli which he makes use of in building a self. The response which he has a tendency to make to these stimuli organizes them. He plays that he is, for instance, offering himself something, and he buys it; he gives a letter to himself and takes it away; he addresses himself as a parent, as a teacher; he arrests himself as a policeman. He has a set of stimuli which call out in himself the sort of responses they call out in others. He takes this group of responses and organizes them into a certain whole. Such is the simplest form of being another to one's self. It involves a temporal situation. The child says something in one character and responds in another character, and then his responding in another character is a stimulus to himself in the first character, and so the conversation goes on. A certain organized structure arises in him and in his other which replies to it, and these carry on the conversation of gestures between themselves.

If we contrast play with the situation in an organized game, we note the essential difference that the child who plays in a game must be ready to take the attitude of everyone else involved in that game, and that these different roles must have definite relationship to each other. Taking a very simple game such as hide-and-seek, everyone with the exception of the one who is hiding is a person who is hunting. A child does not require more than the person who is hunted and the one who is hunting. If a child is playing in the first sense he just goes on playing, but there is no basic organization gained. In that early stage he passes from one to another just as a whim takes him. But in a game where a number of individuals are involved, then the child taking one role must be ready to take the role of everyone else. If he gets in a ball nine he must have the responses of each position involved in his own position. He must know what everyone else is going to do in order to carry out his own play. He has to take all of these roles. They do not all have to be present in consciousness at the same time, but at some moments he has to have three or four individuals present in his own attitude, such as the one who is going to throw the ball, the one who is going to catch it, and so on. These responses must be, in some degree, present in his own makeup. In the game, then, there is a set of responses of such others so organized that the attitude of one calls out the appropriate attitudes of the other.

This organization is put in the form of the rules of the game. Children take a great interest in rules. They make rules on the spot in order to help themselves out of difficulties. Part of the enjoyment of the game is to get these rules. Now, the rules are the set of responses which a particular attitude calls out. You can demand a certain response in others if you take a certain attitude. These responses are all in yourself as well. There you get an organized set of such responses as that to which I have referred, which is something more elaborate than the roles found in play. Here there is just a set of responses that follow on each other indefinitely. At such a stage we speak of a child as not yet having a fully developed self. The child responds in a fairly intelligent fashion to the immediate stimuli that come to him, but they are not organized. He does not organize his life as we would like to have him do, namely, as a whole. There is just a set of responses

of the type of play. The child reacts to a certain stimulus, and the reaction is in himself that is called out in others, but he is not a whole self. In his game he has to have an organization of these roles; otherwise he cannot play the game. The game represents the passage in the life of the child from taking the role of others in play to the organized part that is essential to self-consciousness in the full sense of the term.

. . . The fundamental difference between the game and play is that in the former the child must have the attitude of all the others involved in that game. The attitudes of the other players which the participant assumes organize into a sort of unit, and it is that organization which controls the response of the individual. The illustration used was of a person playing baseball. Each one of his own acts is determined by his assumption of the action of the others who are playing the game. What he does is controlled by his being everyone else on that team, at least insofar as those attitudes affect his own particular response. We get then an "other" which is an organization of the attitudes of those involved in the same process.

The organized community or social group which gives to the individual his unity of self may be called "the generalized other." The attitude of the generalized other is the attitude of the whole community.[4] Thus, for example, in the case of such a social group as a ball team, the team is the generalized other insofar as it enters—as an organized process or social activity—into the experience of any one of the individual members of it.

If the given human individual is to develop a self in the fullest sense, it is not sufficient for him merely to take the attitudes of other human individuals toward himself and toward one another within the human social process, and to bring that social process as a whole into his individual experience merely in these terms: He must also, in the same way that he takes the attitudes of other individuals toward himself and toward one another, take their attitudes toward the various phases or aspects of the common social activity or set of social undertakings in which, as members of an organized society or social group, they are all engaged; and he must then, by generalizing these individual attitudes of that organized society or social group itself, as a whole, act toward different social projects which at any given time it is carrying out, or toward the various larger phases of the general social process which constitutes its life and of which these projects are specific manifestations. This getting of the broad activities of any given social whole or organized society as such within the experiential field of any one of the individuals involved or included in that whole is, in other words, the essential basis and prerequisite of the fullest development of that individual's self: Only insofar as he takes the attitudes of the organized social group to which he belongs toward the organized, cooperative social activity or set of such activities in which that group as such is engaged, does he develop a complete self or possess the sort of complete self he has developed. And on the other hand, the complex cooperative processes and activities and institutional functionings of organized human society are also possible only insofar as every individual involved in them or belonging to that society can take the general attitudes of all other such individuals with reference to these processes and activities and institutional functionings, and to the organized social whole of experiential relations and interactions thereby constituted—and can direct his own behavior accordingly.

It is in the form of the generalized other that the social process influences the behavior of the individuals involved in it and carrying it on, i.e., that the community exercises control over the conduct of its individual members; for it is in this form that the social process or community enters as a determining factor into the individual's thinking. In abstract thought the individual takes the attitude of the generalized other[5] toward himself, without reference to its expression in any particular other individuals; and in concrete thought he takes that attitude insofar as it expresses in the attitudes toward his behavior of those other individuals with whom he is involved in the given social situation or act. But only by taking the attitude of the generalized other toward himself, in one or another of these ways, can he think at all; for only thus can thinking—

or the internalized conversation of gestures which constitutes thinking—occur. And only through the taking by individuals of the attitude or attitudes of the generalized other toward themselves is the existence of a universe of discourse, as that system of common or social meanings which thinking presupposes at its context, rendered possible.

. . . I have pointed out, then, that there are two general stages in the full development of the self. At the first of these stages, the individual's self is considered simply by an organization of the particular attitudes of other individuals toward himself and toward one another in the specific social acts in which he participates with them. But at the second stage in the full development of the individual's self that self is constituted not only by an organization of these particular individual attitudes, but also by an organization of the social attitudes of the generalized other or the social group as a whole to which he belongs. . . . So the self reaches its full development by organizing these individual attitudes of others into the organized social or group attitudes, and by thus becoming an individual reflection of the general systematic pattern of social or group behavior in which it and the others are all involved—a pattern which enters as a whole into the individual's experience in terms of these organized group attitudes which, through the mechanism of his central nervous system, he takes toward himself, just as he takes the individual attitudes of others.

. . . A person is a personality because he belongs to a community, because he takes over the institutions of that community into his own conduct. He takes its language as a medium by which he gets his personality, and then through a process of taking the different roles that all the others furnish he comes to get the attitude of the members of the community. Such, in a certain sense, is the structure of a man's personality. There are certain common responses which each individual has toward certain common things, and insofar as those common responses are awakened in the individual when he is affecting other persons he arouses his own self. The structure, then, on which the self is built is this response which is common to all, for one has to be a member of a community to be a self. Such responses are abstract attitudes, but they constitute just what we term a man's character. They give him what we term his principles, the acknowledged attitudes of all members of the community toward what are the values of that community. He is putting himself in the place of the generalized other, which represents the organized responses of all the members of the group. It is that which guides conduct controlled by principles, and a person who has such an organized group of responses is a man whom we say has character, in the moral sense.

. . . I have so far emphasized what I have called the structures upon which the self is constructed, the framework of the self, as it were. Of course we are not only what is common to all: Each one of the selves is different from everyone else; but there has to be such a common structure as I have sketched in order that we may be members of a community at all. We cannot be ourselves unless we are also members in whom there is a community of attitudes which control the attitudes of all. We cannot have rights unless we have common attitudes. That which we have acquired as self-conscious persons makes us such members of society and gives us selves. Selves can only exist in definite relationships to other selves. No hard-and-fast line can be drawn between our own selves and the selves of others, since our own selves exist and enter as such into our experience only insofar as the selves of others exist and enter as such into our experience also. The individual possesses a self only in relation to the selves of the other members of his social group; and the structure of his self expresses or reflects the general behavior pattern of this social group to which he belongs, just as does the structure of the self of every other individual belonging to this social group.

NOTES

1. Man's behavior is such in his social group that he is able to become an object to himself, a fact which constitutes him a more advanced product of evolutionary development than are the lower animals. Fundamentally it is this social fact—and not his alleged possession of a soul or mind with which he, as an individual, has been mysteriously and supernaturally endowed, and with which the lower animals have not been endowed—that differentiates him from them.

2. It is generally recognized that the specifically social expressions of intelligence, or the exercise of what is often called "social intelligence," depend upon the given individual's ability to take the roles of, or "put himself in the place of," the other individuals implicated with him in given social situations; and upon his consequent sensitivity to their attitudes toward himself and toward one another. These specifically social expressions of intelligence, of course, acquire unique significance in terms of our view that the whole nature of intelligence is social to the very core—that this putting of one's self in the places of others, this taking by one's self of their roles or attitudes, is not merely one of the various aspects or expressions of intelligence or intelligent behavior, but is the very essence of its character. Spearman's "X factor" in intelligence—the unknown factor which, according to him, intelligence contains—is simply (if our social theory of intelligence is correct) this ability of the intelligent individual to take the attitude of the other, or the attitudes of others, thus realizing the significations or grasping the meanings of the symbols or gestures in terms of which thinking proceeds; and thus being able to carry on with himself the internal conversation with these symbols or gestures which thinking involves.

3. The unity of the mind is not identical with the unity of the self. The unity of the self is constituted by the unity of the entire relational pattern of social behavior and experience in which the individual is implicated, and which is reflected in the structure of the self; but many of the aspects or features of this entire pattern do not enter into consciousness, so that the unity of the mind is in a sense an abstraction from the more inclusive unity of the self.

4. It is possible for inanimate objects, no less than for other human organisms, to form parts of the generalized and organized—the completely socialized—other for any given human individual, insofar as he responds to such objects socially or in a social fashion (by means of the mechanism of thought, the internalized conversation of gestures). Any thing—any object or set of objects, whether animate or

inanimate, human or animal, or merely physical—toward which he acts, or to which he responds, socially, is an element in what for him is the generalized other, by taking the attitudes of which toward himself he becomes conscious of himself as an object or individual, and thus develops a self or personality. Thus, for example, the cult, in its primitive form, is merely the social embodiment of the relation between the given social group or community and its physical environment—an organized social means, adopted by the individual members of that group or community, of entering into social relations with that environment, or (in a sense) of carrying on conversations with it; and in this way that environment becomes part of the total generalized other for each of the individual members of the given social group or community.

5. We have said that the internal conversation of the individual with himself in terms of words or significant gestures—the conversation which constitutes the process or activity of thinking—is carried on by the individual from the standpoint of the "generalized other." And the more abstract that conversation is, the more abstract thinking happens to be, the further removed is the generalized other from any connection with particular individuals. It is especially in abstract thinking, that is to say, that the conversation involved is carried on by the individual with the generalized other, rather than with any particular individuals. Thus it is, for example, that abstract concepts are concepts stated in terms of the attitudes of the entire social group or community; they are stated on the basis of the individual's consciousness of the attitudes of the generalized other toward them, as a result of his taking these attitudes of the generalized other and then responding to them. And thus it is also that abstract propositions are stated in a form which anyone—any other intelligent individual—will accept.

DISCUSSION QUESTIONS

1. Mead shows how children's play and games are fundamental in learning the "self." What play and games influenced the formation of your self-concept? How does this vary for different social groups (i.e., groups distinguished by gender, race, class, age, etc.)?

2. What role does reflection have in the formation of the self? Why, according to Mead, is this fundamentally a social process?

3. How does Mead distinguish the self and the body? Taking his analysis further, how do people's reflections on the body influence the formation of self? What other social factors are involved in this process, and what do they tell you about the relationship between the individual and society?

INTERNET RESOURCES

Suggested Web URLs
for Further Study

http://spartan.ac.brocku.ca/%7Elward/
The only site you need for George Herbert Mead. Contains related works, analyses, research and discussion forums.

InfoTrac College Edition

You can find further relevant readings on the World Wide Web at
http://sociology.wadsworth.com

Virtual Society

For further information on this subject including links to relevant Web sites, go to the Wadsworth Sociology homepage at
http://sociology.wadsworth.com

8

The Presentation of Self

ERVING GOFFMAN

When an individual enters the presence of others, they commonly seek to acquire information about him or to bring into play information about him already possessed. They will be interested in his general socioeconomic status, his conception of self, his attitude toward them, his competence, his trustworthiness, etc. Although some of this information seems to be sought almost as an end in itself, there are usually quite practical reasons for acquiring it. Information about the individual helps to define the situation, enabling others to know in advance what he will expect of them and what they may expect of him. Informed in these ways, the others will know how best to act in order to call forth a desired response from him.

For those present, many sources of information become accessible and many carriers (or "sign-vehicles") become available for conveying this information. If unacquainted with the individual, observers can glean clues from his conduct and appearance which allow them to apply their previous experience with individuals roughly similar to the one before them or, more important, to apply untested stereotypes to him. They can also assume from past experience that only individuals of a particular kind are likely to be found in a given social setting. They can rely on what the individual says about himself or on documentary evidence he provides as to who and what he is. If they know, or know of, the individual by virtue of experience prior to the interaction, they can rely on assumptions as to the persistence and generality of psychological traits as a means of predicting his present and future behavior.

However, during the period in which the individual is in the immediate presence of the others, few events may occur which directly provide the others with the conclusive information they will need if they are to direct wisely their own activity. Many crucial facts lie beyond the time and place of interaction or lie concealed within it. For example, the "true" or "real" attitudes, beliefs, and emotions of the individual can be ascertained only indirectly, through his avowals or through what appears to be involuntary expressive behavior. Similarly, if the individual offers the others a product or service, they will often find that during the interaction there will be no time and place immediately available for eating the pudding that the proof can be found in. They will be forced to accept some events as conventional or natural signs of something not directly available to the senses. In Ichheiser's terms,[1] the individual will have to act so that he intentionally or unintentionally *expresses* himself, and the others will in turn have to be *impressed* in some way by him.

The expressiveness of the individual (and therefore his capacity to give impressions) appears to involve two radically different kinds of sign activity: the expression that he *gives,* and the expression that he *gives off.* The first involves verbal symbols or their substitutes which he uses admittedly and solely to convey the information that he and the others are known to attach to these symbols. This is communication in the traditional and narrow sense. The second involves a wide range of action that others can treat as symptomatic of the actor, the expectation being that the action was performed for reasons other than the information conveyed in this way. As we shall have to see, this distinction has an only initial validity. The individual does of course intentionally convey misinformation by means of both of these types of communication, the first involving deceit, the second feigning.

. . . Let us now turn from the others to the point of view of the individual who presents himself before them. He may wish them to think highly of him, or to think that he thinks highly of them, or to perceive how in fact he feels toward them, or to obtain no clear-cut impression; he may wish to ensure sufficient harmony so that the interaction can be sustained, or to defraud, get rid of, confuse, mislead, antagonize, or insult them. Regardless of the particular objective which the individual has in mind and of his motive for having this objective, it will be in his interests to control the conduct of the others, especially their responsive treatment of him. This control is achieved largely by influencing the definition of the situation which the others come to formulate, and he can influence this definition by expressing himself in such a way as to give them the kind of impression that will lead them to act

voluntarily in accordance with his own plan. Thus, when an individual appears in the presence of others, there will usually be some reason for him to mobilize his activity so that it will convey an impression to others which it is in his interests to convey. Since a girl's dormitory mates will glean evidence of her popularity from the calls she receives on the phone, we can suspect that some girls will arrange for calls to be made, and Willard Waller's finding can be anticipated:

> It has been reported by many observers that a girl who is called to the telephone in the dormitories will often allow herself to be called several times, in order to give all the other girls ample opportunity to hear her paged.[2]

Of the two kinds of communication—expressions given and expressions given off—this report will be primarily concerned with the latter, with the more theatrical and contextual kind, the nonverbal, presumably unintentional kind, whether this communication be purposely engineered or not. As an example of what we must try to examine, I would like to cite at length a novelistic incident in which Preedy, a vacationing Englishman, makes his first appearance on the beach of his summer hotel in Spain:

> But in any case he took care to avoid catching anyone's eye. First of all, he had to make it clear to those potential companions of his holiday that they were of no concern to him whatsoever. He stared through them, round them, over them—eyes lost in space. The beach might have been empty. If by chance a ball was thrown his way, he looked surprised; then let a smile of amusement lighten his face (Kindly Preedy), looked round dazed to see that there *were* people on the beach, tossed it back with a smile to himself and not a smile at the people, and then resumed carelessly his nonchalant survey of space.
>
> But it was time to institute a little parade, the parade of the Ideal Preedy. By devious handlings he gave any who wanted to look a

chance to see the title of his book—a Spanish translation of Homer, classic thus, but not daring, cosmopolitan too—and then gathered together his beach-wrap and bag into a neat sand-resistant pile (Methodical and Sensible Preedy), rose slowly to stretch at ease his huge frame (Big-Cat Preedy), and tossed aside his sandals (Carefree Preedy, after all).

The marriage of Preedy and the sea! There were alternative rituals. The first involved the stroll that turns into a run and a dive straight into the water, thereafter smoothing into a strong splashless crawl towards the horizon. But of course not really to the horizon. Quite suddenly he would turn on to his back and thrash great white splashes with his legs, somehow thus showing that he could have swum further had he wanted to, and then would stand up a quarter out of water for all to see who it was.

The alternative course was simpler, it avoided the cold-water shock and it avoided the risk of appearing too high-spirited. The point was to appear to be so used to the sea, the Mediterranean, and this particular beach, that one might as well be in the sea as out of it. It involved a slow stroll down and into the edge of the water—not even noticing his toes were wet, land and water all the same to *him!*—with his eyes up at the sky gravely surveying portents, invisible to others, of the weather (Local Fisherman Preedy).[3]

The novelist means us to see that Preedy is improperly concerned with the extensive impressions he feels his sheer bodily action is giving off to those around him. We can malign Preedy further by assuming that he has acted merely in order to give a particular impression, that this is a false impression, and that the others present receive either no impression at all, or, worse still, the impression that Preedy is affectedly trying to cause them to receive this particular impression. But the important point for us here is that the kind of impression Preedy thinks he is making is

in fact the kind of impression that others correctly and incorrectly glean from someone in their midst. . . .

There is one aspect of the others' response that bears special comment here. Knowing that the individual is likely to present himself in a light that is favorable to him, the others may divide what they witness into two parts; a part that is relatively easy for the individual to manipulate at will, being chiefly his verbal assertions, and a part in regard to which he seems to have little concern or control, being chiefly derived from the expressions he gives off. The others may then use what are considered to be the ungovernable aspects of his expressive behavior as a check upon the validity of what is conveyed by the governable aspects. In this a fundamental asymmetry is demonstrated in the communication process, the individual presumably being aware of only one stream of his communication, the witnesses of this stream and one other. For example, in Shetland Isle one crofter's wife, in serving native dishes to a visitor from the mainland of Britain, would listen with a polite smile to his polite claims of liking what he was eating; at the same time she would take note of the rather rapidity with which the visitor lifted his fork or spoon to his mouth, the eagerness with which he passed food into his mouth, and the gusto expressed in chewing the food, using these signs as a check on the stated feelings of the eater. The same woman, in order to discover what one acquaintance (A) "actually" thought of another acquaintance (B), would wait until B was in the presence of A but engaged in conversation with still another person (C). She would then covertly examine the facial expressions of A as he regarded B in conversation with C. Not being in conversation with B, and not being directly observed by him, A would sometimes relax usual constraints and tactful deceptions, and freely express what he was "actually" feeling about B. This Shetlander, in short, would observe the unobserved observer.

Now given the fact that others are likely to check up on the more controllable aspects of behavior by means of the less controllable, one can expect that sometimes the individual will try to exploit this very possibility, guiding the impression he makes through behavior felt to be reliably informing.[4] For example, in gaining admission to a tight social circle, the participant observer may not only wear an accepting look while listening to an informant, but may also be careful to wear the same look when observing the informant talking to others; observers of the observer will then not as easily discover where he actually stands. A specific illustration may be cited from Shetland Isle. When a neighbor dropped in to have a cup of tea, he would ordinarily wear at least a hint of an expectant warm smile as he passed through the door into the cottage. Since lack of physical obstructions outside the cottage and lack of light within it usually made it possible to observe the visitor unobserved as he approached the house, islanders sometimes took pleasure in watching the visitor drop whatever expression he was manifesting and replace it with a sociable one just before reaching the door. However, some visitors, in appreciating that this examination was occurring, would blindly adopt a social face a long distance from the house, thus ensuring the projection of a constant image.

This kind of control upon the part of the individual reinstates the symmetry of the communication process, and sets the stage for a kind of information game—a potentially infinite cycle of concealment, discovery, false revelation, and rediscovery. It should be added that since the others are likely to be relatively unsuspicious of the presumably unguided aspects of the individual's conduct, he can gain much by controlling it. The others of course may sense that the individual is manipulating the presumably spontaneous aspects of his behavior, and seek in this very act of manipulation some shading of conduct that the individual has not managed to control. This again provides a check upon the individual's behavior, this time his presumably uncalculated behavior, thus re-establishing the asymmetry of the communication process. Here I would like only to add the suggestion that the arts of piercing an individual's effort at calculated unintentionality seem better developed than our capacity to manipulate our own behavior, so that regardless of how many steps have occurred in the information game, the witness is likely to have the advantage over the actor, and the initial asymmetry of the communication process is likely to be retained. . . .

In everyday life, of course, there is a clear understanding that first impressions are important. Thus, the work adjustment of those in service occupations will often hinge upon a capacity to seize and hold the initiative in the service relation, a capacity that will require subtle aggressiveness on the part of the server when he is of lower socio-economic status than his client. W. F. Whyte suggests the waitress as an example:

> The first point that stands out is that the waitress who bears up under pressure does not simply respond to her customers. She acts with some skill to control their behavior. The first question to ask when we look at the customer relationship is, "Does the waitress get the jump on the customer, or does the customer get the jump on the waitress?" The skilled waitress realizes the crucial nature of this question. . . .
>
> The skilled waitress tackles the customer with confidence and without hesitation. For example, she may find that a new customer has seated himself before she could clear off the dirty dishes and change the cloth. He is now leaning on the table studying the menu. She greets him, says, "May I change the cover, please?" and, without waiting for an answer, takes his menu away from him so that he moves back from the table, and she goes about her work. The relationship is handled politely but firmly, and there is never any question as to who is in charge.[5]

When the interaction that is initiated by "first impressions" is itself merely the initial interaction in an extended series of interactions involving the same participants, we speak of "getting off on the right foot" and feel that it is crucial that we do so. Thus, one learns that some teachers take the following view:

> You can't ever let them get the upper hand on you or you're through. So I start out tough. The first day I get a new class in, I let them know who's boss. . . . You've got to start off tough, then you can ease up as you go along. If you start out easy-going, when you try to get tough, they'll just look at you and laugh.[6]

. . . In stressing the fact that the initial definition of the situation projected by an individual tends to provide a plan for the cooperative activity that follows—in stressing this action point of view— we must not overlook the crucial fact that any projected definition of the situation also has a distinctive moral character. It is this moral character of projections that will chiefly concern us in this report. Society is organized on the principle that any individual who possesses certain social characteristics has a moral right to expect that others will value and treat him in an appropriate way. Connected with this principle is a second, namely that an individual who implicitly or explicitly signifies that he has certain social characteristics ought in fact to be what he claims he is. In consequence, when an individual projects a definition of the situation and thereby makes an implicit or explicit claim to be a person of a particular kind, he automatically exerts a moral demand upon the others, obliging them to value and treat him in the manner that persons of his kind have a right to expect. He also implicitly foregoes all claims to be things he does not appear to be[7] and hence foregoes the treatment that would be appropriate for such individuals. The others find, then, that the individual has informed them as to what is and as to what they *ought* to see as the "is."

One cannot judge the importance of definitional disruptions by the frequency with which they occur, for apparently they would occur more frequently were not constant precautions taken. We find that preventive practices are constantly employed to avoid these embarrassments and that corrective practices are constantly employed to compensate for discrediting occurrences that have not been successfully avoided. When the individual employs these strategies and tactics to protect his own projections, we may refer to them as "defensive practices"; when a participant employs them to save the definition of the situation projected by another, we speak of "protective practices" or "tact." Together, defensive and protective practices comprise the techniques employed to safeguard the impression fostered by an individual during his presence before others. It should be added that while we may be ready to see that no fostered impression would survive if defensive practices were not employed, we are less ready perhaps to see that few impressions could survive if those who received the impression did not exert tact in their reception of it.

In addition to the fact that precautions are taken to prevent disruption of projected definitions, we may also note that an intense interest in these disruptions comes to play a significant role in the social life of the group. Practical jokes and social games are played in which embarrassments which are to be taken unseriously are purposely engineered.[8] Fantasies are created in which devastating exposures occur. Anecdotes from the past—real, embroidered, or fictitious—are told and retold, detailing disruptions which occurred, almost occurred, or occurred and were admirably resolved. There seems to be no grouping which does not have a ready supply of these games, reveries, and cautionary tales, to be used as a source of humor, a catharsis for anxieties, and a sanction for inducing individuals to be modest in their claims and reasonable in their projected expectations. The individual may tell himself through dreams of getting into impossible positions. Families tell of the time a guest got his dates mixed and arrived when neither the house nor anyone in it was ready for him. Journalists tell of times when an all too-meaningful misprint occurred, and the paper's assumption of objectivity or decorum was humorously discredited. Public servants tell of times a client ridiculously misunderstood form instructions, giving answers which implied an unanticipated and bizarre definition of the situation.[9] Seamen, whose home away from home is rigorously he-man, tell stories of coming back home and inadvertently asking mother to "pass the fucking butter."[10] Diplomats tell of the time a near-sighted queen asked a republican ambassador about the health of his king.[11]

To summarize, then, I assume that when an individual appears before others he will have many motives for trying to control the impression they receive of the situation.

NOTES

1. Gustav Ichheiser, "Misunderstandings in Human Relations," supplement to *The American Journal of Sociology*, 55 (Sept. 1949), 6–7.

2. Willard Waller, "The Rating and Dating Complex," *American Sociological Review*, 2, 730.

3. William Sansom, *A Contest of Ladies* (London: Hogarth, 1956), pp. 230–31.

4. The widely read and rather sound writings of Stephen Potter are concerned in part with signs that can be engineered to give a shrewd observer the apparently incidental cues he needs to discover concealed virtues the gamesman does not in fact possess.

5. W. F. Whyte, "When Workers and Customers Meet," chap. 7, *Industry and Society*, ed. W. F. Whyte (New York: McGraw-Hill, 1946), pp. 132–33.

6. Teacher interview quoted by Howard S. Becker, "Social Class Variations in the Teacher–Pupil Relationship," *Journal of Educational Sociology*, 25, 459.

7. This role of the witness in limiting what it is the individual can be has been stressed by Existentialists, who see it as a basic threat to individual freedom. See Jean-Paul Sartre, *Being and Nothingness*, trans. Hazel E. Barnes (New York: Philosophical Library, 1956), pp. 364ff.

8. Goffman, op. cit., pp. 319–27.

9. Peter Blau, "Dynamics of Bureaucracy" (Ph.D. dissertation, Department of Sociology, Columbia University, forthcoming, University of Chicago Press), pp. 127–29.

10. Walter M. Beattie, Jr., "The Merchant Seaman" (unpublished M.A. Report, Department of Sociology, University of Chicago, 1950), p. 35.

11. Sir Frederick Ponsonby, *Recollections of Three Reigns* (New York: Dutton, 1952), p. 46.

DISCUSSION QUESTIONS

1. Think of a time when you were particularly concerned about making an impression on a group or person previously unknown to you. What did you do to give cues so that you could control the other's impression of you? How did you know what "cues" to give? How does this support Goffman's argument about the presentation of self?

2. Have you ever been in a situation in which your presentation of self failed? What happened? How does this illustrate Goffman's point that precautions are usually taken to prevent disruption of a projected definition?

3. Have you ever had a time when you had to give a presentation of self that you felt was "not you"? What happened? How does this illustrate Goffman's point that an individual who signifies certain social characteristics is assumed to be what he or she claims to be?

INTERNET RESOURCES

Suggested Web URLs
for Further Study

http://www.cfmc.com/adamb/writings/goffman.htm
This site is both critic of and helpful guide through the concepts presented in *The Presentation of Self.*

http://itsa.ucsf.edu/~eliotf/Celebrating_Erving_Goffman.html
A post-mortem analysis of Erving Goffman through his two best-known early works, *The Presentation of Self* and *Asylum.*

InfoTrac College Edition

You can find further relevant readings on the World Wide Web at
http://sociology.wadsworth.com

Virtual Society

For further information on this subject including links to relevant Web sites, go to the Wadsworth Sociology homepage at
http://sociology.wadsworth.com

9

The Uses of Poverty:
The Poor Pay All

HERBERT J. GANS

Some years ago Robert K. Merton applied the notion of functional analysis to explain the continuing though maligned existence of the urban political machine: If it continued to exist, perhaps it fulfilled latent—unintended or unrecognized—positive functions. Clearly it did. Merton pointed out how the political machine provided central authority to get things done when a decentralized local government could not act, humanized the services of the impersonal bureaucracy for fearful citizens, offered concrete help (rather than abstract law or justice) to the poor, and otherwise performed services needed or demanded by many people but considered unconventional or even illegal by formal public agencies.

Today, poverty is more maligned than the political machine ever was; yet it, too, is a persistent social phenomenon. Consequently, there may be some merit in applying functional analysis to poverty, in asking whether it also has positive functions that explain its persistence.

Merton defined functions as "those observed consequences [of a phenomenon] which make for the adaptation of adjustment of a given [social] system." I shall use a slightly different definition; instead of identifying functions for an entire social system, I shall identify them for the interest groups, socio-economic classes, and other population aggregates with shared values that "inhabit" a social system. I suspect that in a modern heterogeneous society, few phenomena are functional or dysfunctional for the society as a whole, and that most result in benefits to some groups and costs to others. Nor are any phenomena indispensable; in most instances, one can suggest what Merton calls "functional alternatives" or equivalents for them, i.e., other social patterns or policies that achieve the same positive functions but avoid the dysfunction. (In the following discussion, positive functions will be abbreviated as functions and negative functions as dysfunctions. Functions and dysfunctions, in the planner's terminology, will be described as benefits and costs.)

Associating poverty with positive functions seems at first glance to be unimaginable. Of course, the slumlord and the loan shark are commonly known to profit from the existence of poverty, but they are viewed as evil men, so their activities are classified among the dysfunctions of poverty. However, what is less often recognized, at least by the conventional wisdom, is that poverty also makes possible the existence or expansion

Credit: "The Uses of Poverty," by Herbert Gans from *Social Policy*, July/Aug. 1971, pp. 20–24, ©1971 by Social Policy Corp.

of respectable professions and occupations, for example, penology, criminology, social work, and public health. More recently, the poor have provided jobs for professional and para-professional "poverty warriors," and for journalists and social scientists, this author included, who have supplied the information demanded by the revival of public interest in poverty.

Clearly, then, poverty and the poor may well satisfy a number of positive functions for many nonpoor groups in American society. I shall describe 13 such functions—economic, social, and political—that seem to me most significant.

THE FUNCTIONS OF POVERTY

First, the existence of poverty ensures that society's "dirty work" will be done. Every society has such work: physically dirty or dangerous, temporary, dead-end and underpaid, undignified, and menial jobs. Society can fill these jobs by paying higher wages than for "clean" work, or it can force people who have no other choice to do the dirty work—and at low wages. In America, poverty functions to provide a low-wage labor pool that is willing—or, rather, unable to be *un*willing—to perform dirty work at low cost. Indeed, this function of the poor is so important that in some Southern states, welfare payments have been cut off during the summer months when the poor are needed to work in the field. Moreover, much of the debate about the Negative Income Tax and the Family Assistance Plan has concerned their impact on the work incentive, by which is actually meant the incentive of the poor to do the needed dirty work if the wages therefrom are no larger than the income grant. Many economic activities that involve dirty work depend on the poor for their existence: restaurants, hospitals, parts of the garment industry, and "truck farming," among others, could not persist in their present form without the poor.

Second, because the poor are required to work at low wages, they subsidize a variety of economic activities that benefit the affluent. For example, domestics subsidize the upper-middle and upper classes, making life easier for their employers and freeing affluent women for a variety of professional, cultural, civic, and partying activities. Similarly, because the poor pay a higher proportion of their income in property and sales taxes, among others, they subsidize many state and local governmental services that benefit more affluent groups. In addition, the poor support innovation in medical practice as patients in teaching and research hospitals and as guinea pigs in medical experiments.

Third, poverty creates jobs for a number of occupations and professions that serve or "service" the poor, or protect the rest of society from them. As already noted, penology would be minuscule without the poor, as would the police. Other activities and groups that flourish because of the existence of poverty are the numbers game, the sale of heroin and cheap wines and liquors, pentecostal ministers, faith healers, prostitutes, pawn shops, and the peacetime army, which recruits its enlisted men mainly from among the poor.

Fourth, the poor buy goods others do not want and thus prolong the economic usefulness of such goods—day-old bread, fruit and vegetables that would otherwise have to be thrown out, secondhand clothes, and deteriorating automobiles and buildings. They also provide incomes for doctors, lawyers, teachers and others who are too old, poorly trained, or incompetent to attract more affluent clients.

In addition to economic functions, the poor perform a number of social functions.

Fifth, the poor can be identified and punished as alleged or real deviants in order to uphold the legitimacy of conventional norms. To justify the desirability of hard work, thrift, honesty, and monogamy, for example, the defenders of these norms must be able to find people who can be accused of being lazy, spendthrift, dishonest, and promiscuous. Although there is some evidence that the poor are about as moral and law-abiding as anyone else, they are more likely than middle-class transgressors to be caught and punished when they participate in deviant acts. Moreover, they lack the political and cultural power to correct the stereotypes that other people hold of them and thus continue to be thought of as lazy, spendthrift, etc., by those who need living proof that moral deviance does not pay.

Sixth, and conversely, the poor offer vicarious participation to the rest of the population in the uninhibited sexual, alcoholic, and narcotic behavior in which they are alleged to participate and which, being freed from the constraints of affluence, they are often thought to enjoy more than the middle classes. Thus many people, some social scientists included, believe that the poor not only are more given to uninhibited behavior (which may be true, although it is often motivated by despair more than by lack of inhibition) but derive more pleasure from it than affluent people (which research by Lee Rainwater, Walter Miller, and others shows to be patently untrue). However, whether the poor actually have more sex and enjoy it more is irrelevant; so long as middle-class people believe this to be true, they can participate in it vicariously when instances are reported in factual or fictional form.

Seventh, the poor also serve a direct cultural function when culture created by or for them is adopted by the more affluent. The rich often collect artifacts from extinct folk cultures of poor people; and almost all Americans listen to the blues, Negro spirituals, and country music, which originated among the Southern poor. Recently they have enjoyed the rock styles that were born, like the Beatles, in the slums; and in the last year, poetry written by ghetto children has become popular in literary circles. The poor also serve as culture heroes, particularly, of course, to the left; but the hobo, the cowboy, the hipster, and the mythical prostitute with a heart of gold have performed this function for a variety of groups.

Eighth, poverty helps to guarantee the status of those who are not poor. In every hierarchical society someone has to be at the bottom; but in American society, in which social mobility is an important goal for many and people need to know where they stand, the poor function as a reliable and relatively permanent measuring rod for status comparisons. This is particularly true for the working class, whose politics is influenced by the need to maintain status distinctions between themselves and the poor, much as the aristocracy must find ways of distinguishing itself from the *nouveaux riches*.

Ninth, the poor also aid the upward mobility of groups just above them in the class hierarchy. Thus a goodly number of Americans have entered the middle class through the profits earned from the provision of goods and services in the slums, including illegal or nonrespectable ones that upper-class and upper-middle-class businessmen shun because of their low prestige. As a result, members of almost every immigrant group have financed their upward mobility by providing slum housing, entertainment, gambling, narcotics, etc., to later arrivals—most recently to blacks and Puerto Ricans.

Tenth, the poor help to keep the aristocracy busy, thus justifying its continued existence. "Society" uses the poor as clients of settlement houses and beneficiaries of charity affairs; indeed, the aristocracy must have the poor to demonstrate its superiority over other elites who devote themselves to earning money.

Eleventh, the poor, being powerless, can be made to absorb the costs of change and growth in American society. During the nineteenth century, they did the backbreaking work that built the cities; today, they are pushed out of their neighborhoods to make room for "progress." Urban renewal projects to hold middle-class taxpayers in the city and expressways to enable suburbanites to commute downtown have typically been located in poor neighborhoods, since no other group will allow itself to be displaced. For the same reason, universities, hospitals, and civic centers also expand into land occupied by the poor. The major costs of the industrialization of agriculture have been borne by the poor, who are pushed off the land without recompense; and they have paid a large share of the human cost of the growth of American power overseas, for they have provided many of the foot soldiers for Vietnam and other wars.

Twelfth, the poor facilitate and stabilize the American political process. Because they vote and participate in politics less than other groups, the political system is often free to ignore them. Moreover, since they can rarely support Republicans, they often provide the Democrats with a captive constituency that has no other place to go. As a result, the Democrats can count on their votes, and be more responsive to voters—for example, the white working class— who might otherwise switch to the Republicans.

Thirteenth, the role of the poor in upholding conventional norms (see the *fifth* point, above) also has a significant political function. An economy based on the ideology of laissez-faire requires a deprived population that is allegedly unwilling to work or that can be considered inferior because it must accept charity or welfare in order to survive. Not only does the alleged moral deviancy of the poor reduce the moral pressure on the present political economy to eliminate poverty, but socialist alternatives can be made to look quite unattractive if those who will benefit most from them can be described as lazy, spendthrift, dishonest, and promiscuous.

THE ALTERNATIVES

I have described 13 of the more important functions poverty and the poor satisfy in American society, enough to support the functionalist thesis that poverty, like any other social phenomenon, survives in part because it is useful to society or some of its parts. This analysis is not intended to suggest that because it is often functional, poverty *should* exist, or that it *must* exist. For one thing, poverty has many more dysfunctions than functions; for another, it is possible to suggest functional alternatives.

For example, society's dirty work could be done without poverty, either by automation or by

paying "dirty workers" decent wages. Nor is it necessary for the poor to subsidize the many activities they support through their low-wage jobs. This would, however, drive up the costs of these activities, which would result in higher prices to their customers and clients. Similarly, many of the professionals who flourish because of the poor could be given other roles. Social workers could provide counseling to the affluent, as they prefer to do anyway; and the police could devote themselves to traffic and organized crime. Other roles would have to be found for badly trained or incompetent professionals now relegated to serving the poor, and someone else would have to pay their salaries. Fewer penologists would be employable, however. And pentecostal religion could probably not survive without the poor—nor would parts of the second- and third-hand-goods market. And in many cities, "used" housing that no one else wants would then have to be torn down at public expense.

Alternatives for the cultural functions of the poor could be found more easily and cheaply. Indeed, entertainers, hippies, and adolescents are already serving as the deviants needed to uphold traditional morality and as devotees of orgies to "staff" the fantasies of vicarious participation.

The status functions of the poor are another matter. In a hierarchical society, some people must be defined as inferior to everyone else with respect to a variety of attributes, but they need not be poor in the absolute sense. One could conceive of a society in which the "lower class," though last in the pecking order, received 75 percent of the median income, rather than 15–40 percent, as is now the case. Needless to say, this would require considerable income redistribution.

The contribution the poor make to the upward mobility of the groups that provide them with goods and services could also be maintained without the poor's having such low incomes. However, it is true that if the poor were more affluent, they would have access to enough capital to take over the provider role, thus competing with, and perhaps rejecting, the "outsiders." (Indeed, owing in part to antipoverty programs, this is already happening in a number of ghettos, where white storeowners are being replaced by blacks.) Similarly, if the poor were more affluent, they would make less willing clients for upper-class philanthropy, although some would still use settlement houses to achieve upward mobility, as they do now. Thus "Society" could continue to run its philanthropic activities.

The political functions of the poor would be more difficult to replace. With increased affluence the poor would probably obtain more political power and be more active politically. With higher incomes and more political power, the poor would be likely to resist paying the costs of growth and change. Of course, it is possible to imagine urban renewal and highway projects that properly reimbursed the displaced people, but such projects would then become considerably more expensive, and many might never be built. This, in turn, would reduce the comfort and convenience of those who now benefit from urban renewal and expressways. Finally, hippies could serve also as more deviants to justify the existing political economy—as they already do. Presumably, however, if poverty were eliminated, there would be fewer attacks on that economy.

In sum, then, many of the functions served by the poor could be replaced if poverty were eliminated, but almost always at higher costs to others, particularly more affluent others. Consequently, a functional analysis must conclude that poverty persists not only because it fulfills a number of positive functions but also because many of the functional alternatives to poverty would be quite dysfunctional for the affluent members of society. A functional analysis thus ultimately arrives at much the same conclusion as radical sociology, except that radical thinkers treat as manifest what I describe as latent: that social phenomena that are functional for affluent or powerful groups and dysfunctional for poor or powerless ones persist; that when the elimination of such phenomena through functional alternatives would generate dysfunctions for the affluent or powerful, they will continue to persist; and that phenomena like poverty can be eliminated only when they become dysfunctional for the affluent or powerful, or when the powerless can obtain enough power to change society.

POSTSCRIPT

Over the years, this article has been interpreted as either a direct attack on functionalism or a tongue-in-cheek satirical comment on it. Neither interpretation is true. I wrote the article for two reasons. First and foremost, I wanted to point out that there are, unfortunately, positive functions of poverty which have to be dealt with by antipoverty policy. Second, I was trying to show that functionalism is not the inherently conservative approach for which it has often been criticized, but that it can be employed in liberal and radical analyses.

DISCUSSION QUESTIONS

1. What specific groups does Gans identify as benefiting from poverty?

2. Gans wrote this essay in 1971. Were he writing it today, what specific observations about the functions of poverty might he further identify? Be as specific as possible.

3. Gans uses the perspective of functionalist theory to think about poverty; at the same time, he draws from conflict theory. What do each of these perspectives contribute to an understanding of poverty in society?

INTERNET RESOURCES

Suggested Web URLs
for Further Study

http://www.nextcity.com/main/town/4editor.htm #baster
An editorial and responses out of Canada about poverty. A very interesting article bound to spark debate.

http://www.ssc.wisc.edu/irp/
Official site for the Institute for Research on Poverty. IRP is a national nonprofit and nonpartisan organization dedicated to researching the causes and consequences of poverty and social inequality in the U.S.

InfoTrac College Edition

You can find further relevant readings on the World Wide Web at
http://sociology.wadsworth.com

Virtual Society

For further information on this subject including links to relevant Web sites, go to the Wadsworth Sociology homepage at
http://sociology.wadsworth.com

10

The Souls of Black Folk

W.E.B. DUBOIS

Between me and the other world there is ever an unasked question; unasked by some through feelings of delicacy; by others through the difficulty of rightly framing it. All, nevertheless, flutter round it. They approach me in a half-hesitant sort of way, eye me curiously or compassionately, and then, instead of saying directly, How does it feel to be a problem? they say, I know an excellent colored man in my town; or, I fought at Mechanicsville; or, Do not these Southern outrages make your blood boil? At these I smile, or am interested, or reduce the boiling to a simmer, as the occasion may require. To the real question, How does it feel to be a problem? I answer seldom a word.

And yet, being a problem is a strange experience,—peculiar even for one who has never been anything else, save perhaps in babyhood and in Europe. It is in the early days of rollicking boyhood that the revelation first bursts upon one, all in a day, as it were. I remember well when the shadow swept across me. I was a little thing, away up in the hills of New England, where the dark Housatonic winds between Hoosac and Taghkanic to the sea. In a wee wooden schoolhouse, something put it into the boys' and girls' heads to buy gorgeous visiting cards—ten cents a package—and exchange. The exchange was merry, till one girl, a tall newcomer, refused my card—refused it peremptorily, with a glance. Then it dawned upon me with a certain suddenness that I was different from the others; or like, mayhap, in heart and life and longing, but shut out from their world by a vast veil. I had thereafter no desire to tear down that veil, to creep through; I held all beyond it in common contempt, and lived above it in a region of blue sky and great wandering shadows. That sky was bluest when I could beat my mates at examination-time, or beat them at a foot-race, or even beat their stringy heads. Alas, with the years all this fine contempt began to fade; for the words I longed for, and all their dazzling opportunities, were theirs, not mine. But they should not keep these prizes, I said; some, all, I would wrest from them. Just how I would do it I could never decide: by reading law, by healing the sick, by telling the wonderful tales that swam in my head,—some way. With other black boys the strife was not so fiercely sunny: Their youth shrunk into tasteless sycophancy, or into silent hatred of the pale world about them and mocking distrust of everything white; or wasted itself in a bitter cry, Why did God make me an outcast and stranger in mine own house? The shades of the prison-house closed round about us all: walls strait and stubborn to the whitest, but relentlessly narrow, tall, and unscalable to sons of night who must plod darkly on in resignation, or beat unavailing palms against the stone, or steadily, half hopelessly, watch the streak of blue above.

After the Egyptian and Indian, the Greek and Roman, the Teuton and Mongolian, the Negro is a sort of seventh son, born with a veil, and gifted with second-sight in this American world—a world which yields him no true self-consciousness, but only lets him see himself through the revelation of the other world. It is a peculiar sensation, this double-consciousness, this sense of always looking at one's self through the eyes of others, of measuring one's soul by the tape of a world that looks on in amused contempt and pity. One ever feels his twoness—an American, a Negro; two souls, two thoughts, two unreconciled strivings; two warring ideals in one dark body, whose dogged strength alone keeps it from being torn asunder.

The history of the American Negro is the history of this strife—this longing to attain self-conscious manhood, to merge his double self into a better and truer self. In this merging he wishes neither of the older selves to be lost. He would not Africanize America, for America has too much to teach the world and Africa. He would not bleach his Negro soul in a flood of white Americanism, for he knows that Negro blood has a message for the world. He simply wishes to make it possible for a man to be both a Negro and an American, without being cursed and spit upon by his fellows, without having the doors of Opportunity closed roughly in his face.

This, then, is the end of his striving: to be a coworker in the kingdom of culture, to escape both death and isolation, to husband and use his best powers and his latent genius. These powers of body and mind have in the past been strangely wasted, dispersed, or forgotten. The shadow of a mighty Negro past flits through the tale of Ethiopia the Shadowy and of Egypt the Sphinx. Through history, the powers of single black men flash here and there like falling stars, and die sometimes before the world has rightly gauged their brightness. Here in America, in the few days

since Emancipation, the black man's turning hither and thither in hesitant and doubtful striving has often made his very strength to lose effectiveness, to seem like absence of power, like weakness. And yet it is not weakness—it is the contradiction of double aims. The double-aimed struggle of the black artisan on the one hand to escape white contempt for a nation of mere hewers of wood and drawers of water, and on the other hand to plough and nail and dig for a poverty-stricken horde—could only result in making him a poor craftsman, for he had but half a heart in either cause. By the poverty and ignorance of his people, the Negro minister or doctor was tempted toward quackery and demagogy; and by the criticism of the other world, toward ideals that made him ashamed of his lowly tasks. The would-be black *savant* was confronted by the paradox that the knowledge his people needed was a twice-told tale to his white neighbors, while the knowledge which would teach the white world was Greek to his own flesh and blood. The innate love of harmony and beauty that set the ruder souls of his people a-dancing and a-singing raised but confusion and doubt in the soul of the black artist; for the beauty revealed to him was the soul-beauty of a race which his larger audience despised, and he could not articulate the message of another people. This waste of double aims, this seeking to satisfy two unreconciled ideals, has wrought sad havoc with the courage and faith and deeds of ten thousand people—has sent them often wooing false gods and invoking false means of salvation, and at times has even seemed about to make them ashamed of themselves.

Away back in the days of bondage they thought to see in one divine event the end of all doubt and disappointment; few men ever worshipped Freedom with half such unquestioning faith as did the American Negro for two centuries. To him, so far as he thought and dreamed, slavery was indeed the sum of all villainies, the cause of all sorrow, the root of all prejudice; Emancipation was the key to a promised land of sweeter beauty than ever stretched before the eyes of wearied Israelites. In song and exhortation swelled one refrain—Liberty; in his tears and curses the God he implored had Freedom in his right hand. At last it came—suddenly, fearfully, like a dream. With one wild carnival of blood and passion came the message in his own plaintive cadences:

> Shout, O children!
> Shout, you're free!
> For God has bought your liberty!

Years have passed away since then—ten, twenty, forty; forty years of national life, forty years of renewal and development, and yet the swarthy spectre sits in its accustomed seat at the Nation's feast. In vain do we cry to this our vastest social problem:

> Take any shape but that, and my firm nerves shall never tremble!

The Nation has not yet found peace from its sins; the freedman has not yet found in freedom his promised land. Whatever of good may have come in these years of change, the shadow of a deep disappointment rests upon the Negro people—a disappointment all the more bitter because the unattained ideal was unbounded save by the simple ignorance of a lowly people.

The first decade was merely a prolongation of the vain search for freedom, the boon that seemed ever barely to elude their grasp—like a tantalizing will-o'-the-wisp, maddening and misleading the headless host. The holocaust of war, the terrors of the Ku-Klux Klan, the lies of carpet-baggers, the disorganization of industry, and the contradictory advice of friends and foes, left the bewildered serf with no new watchword beyond the old cry for freedom. As the time flew, however, he began to grasp a new idea. The ideal of liberty demanded for its attainment powerful means, and these the Fifteenth Amendment gave him. The ballot, which before he had looked upon as a visible sign of freedom, he now regarded as the chief means of gaining and perfecting the liberty with which war had partially endowed him. And why not? Had not votes made war and emancipated millions? Had not votes enfranchised the freedmen? Was anything impossible to a power that had done all this? A million black men started with renewed zeal to vote themselves into the kingdom. So the decade flew away, the revolution of 1876 came, and left the half-free serf weary, wondering, but still inspired. Slowly but steadily, in the following years, a new vision began gradually to replace the dream of political power—a powerful movement, the rise of another ideal to guide the unguided, another pillar of fire by night after a clouded day. It was the ideal of "book-learning"; the curiosity, born of compulsory ignorance, to know and test the power of the cabalistic letters of the white man, the longing to know. Here at last seemed to have been discovered the mountain path to Canaan; longer than the highway of Emancipation and law, steep and rugged, but straight, leading to heights high enough to overlook life.

Up the new path the advance guard toiled, slowly, heavily, doggedly; only those who have watched and guided the faltering feet, the misty minds, the dull understandings, of the dark pupils of these schools know how faithfully, how piteously, this people strove to learn. It was weary work. The cold statistician wrote down the inches of progress here and there, noted also where here and there a foot had slipped or some one had fallen. To the tired climbers, the horizon was ever dark, the mists were often cold, the Canaan was always dim and far away. If, however, the vistas disclosed as yet no goal, no resting-place, little but flattery and criticism, the journey at least gave leisure for reflection and self-examination; it changed the child of Emancipation to the youth with dawning self-consciousness, self-realization, self-respect. In those sombre forests of his striving his own soul rose before him, and he saw himself—darkly as through a veil; and yet he saw in himself some faint revelation of his power, of his mission. He began to have a dim feeling that, to attain his place in the world, he must be himself, and not another. For the first time he sought to analyze the burden he bore upon his back, that dead-weight of social degradation partially masked behind a half-named Negro problem. He felt his poverty; without a cent, without a home, without land, tools, or savings, he had entered into competition with rich, landed, skilled neighbors. To be a poor man is hard, but to be a poor race in a land of dollars is the very bottom of hardships. He felt the weight of his ignorance—not simply of letters, but of life, of business, of the humanities; the accumulated sloth and shirking and awkwardness of decades and centuries shackled his hands and feet. Nor was his burgeon all poverty and ignorance. The red stain of bastardy, which two centuries of systematic legal defilement of Negro women had stamped upon his race, meant not only the loss of ancient African chastity, but also the hereditary weight of a mass of corruption from white adulterers, threatening almost the obliteration of the Negro home.

A people thus handicapped ought not to be asked to race with the world, but rather allowed to give all its time and thought to its own social problems. But alas! while sociologists gleefully count his bastards and his prostitutes, the very soul of the toiling, sweating black man is darkened by the shadow of a vast despair. Men call the shadow prejudice, and learnedly explain it as the natural defence of culture against barbarism, learning against ignorance, purity against crime, the "higher" against the "lower" races. To which the Negro cries Amen! and swears that to so much of this strange prejudice as is founded on just homage to civilization, culture, righteousness, and progress, he humbly bows and meekly does obeisance. But before that nameless prejudice that leaps beyond all this he stands helpless, dismayed, and well-nigh speechless; before that personal disrespect and mockery, the ridicule and systematic humiliation, the distortion of fact and wanton license of fancy, the cynical ignoring of the better and the boisterous welcoming of the worse, the all-pervading desire to inculcate disdain for everything black, from Toussaint to the devil—before this there rises a sickening despair that would disarm and discourage any nation save that black host to whom "discouragement" is an unwritten word.

But the facing of so vast a prejudice could not but bring the inevitable self-questioning, self-disparagement, and lowering of ideals which ever accompany repression and breed in an atmosphere of contempt and hate. Whisperings and portents came borne upon the four winds: Lo! we are diseased and dying, cried the dark hosts; we cannot write, our voting is vain; what need of education, since we must always cook and serve? And the Nation echoed and enforced this self-criticism saying: Be content to be servants, and nothing more; what need of higher culture for half-men? Away with the black man's ballot, by force or fraud—and behold the suicide of a race! Nevertheless, out of the evil came something of good—the more careful adjustment of education to real life, the clear perception of the Negroes' social responsibilities, and the sobering realization of the meaning of progress.

So dawned the time of *Sturm und Drang*: Storm and stress to-day rocks our little boat on the mad waters of the world-sea; there is within and without the sound of conflict, the burning of body and rending of soul; inspiration strives with doubt, and faith with vain questionings. The bright ideals of the past—physical freedom, political power, the training of brains and the training of hands—all these in turn have waxed and waned, until even the last grows dim and overcast. Are they all wrong—all false? No, not that, but each alone was oversimple and incomplete—the dreams of a credulous race—childhood, or the fond imaginings of the other world which does not know and does not want to know our power. To be really true, all these ideals must be melted and welded into one. The training of the schools we need to-day more than ever—the training of deft hands, quick eyes and ears, and above all the broader, deeper, higher culture of gifted minds and pure hearts. The power of the ballot we need in sheer self-defense—else what shall save us from a second slavery? Freedom, too, the long-sought, we still seek—the freedom of life and climb, the freedom to work and think, the freedom to love and aspire. Work, culture, liberty—all these we need, not singly but together, not successively but together, each growing and aiding each, and all striving toward that vaster ideal that swims before the Negro people, the ideal of human brotherhood, gained through the unifying ideal of Race; the ideal of fostering and developing the traits and talents of the Negro, not in opposition to or contempt for other races, but rather in large conformity to the greater ideals of the American Republic, in order that some day on American soil two world-races may give each to each those characteristics both so sadly lack. We the darker ones come even now not altogether empty-handed: There are to-day no truer exponents of the pure human spirit of the Declaration of Independence than the American Negroes; there is no true American music but the wild sweet melodies of the Negro slave, the American fairy tales and folklore are Indian and African; and, all in all, we black men seem the sole oasis of simple faith and reverence in a dusty desert of dollars and smartness. Will America be poorer if she replace her brutal dyspeptic blundering with lighthearted but determined Negro humility? Or her coarse and cruel wit with loving jovial good-humor? or her vulgar music with the soul of the Sorrow Songs?

Merely a concrete test of the underlying principles of the great republic is the Negro Problem, and the spiritual striving of the freedmen's sons is the travail of souls whose burden is almost beyond the measure of their strength, but who bear it in the name of an historic race, in the name of this the land of their fathers' fathers, and in the name of human opportunity.

DISCUSSION QUESTIONS

1. What are some of the consequences of racial prejudice, according to DuBois? What evidence do you see of this in contemporary race relations?

2. DuBois was writing at the beginning of the 20th century. Suppose he returned to the United States now. Following from his original ideas, what do you think he would write now?

3. DuBois's concept of "the veil" has been used to refer to the double consciousness that African Americans have to have as an oppressed group in society—that is, knowing themselves and knowing the oppressor. Why does this emerge and what implications does it have for understanding how people's knowledge develops in a racially stratified society?

INTERNET RESOURCES

Suggested Web URLs
for Further Study

http://www.lucidcafe.com/library/96feb/dubois.html
This site provides a biography of DuBois, a list of his books and links to other civil rights leaders.

http://eserver.org/race/resources.html
A site dedicated to race and ethnicity with a list of resources, articles, essays, texts, narratives, and more.

InfoTrac College Edition

You can find further relevant readings on the World Wide Web at
http://sociology.wadsworth.com

Virtual Society

For further information on this subject including links to relevant Web sites, go to the Wadsworth Sociology homepage at
http://sociology.wadsworth.com

11

The Future of Marriage

JESSIE BERNARD

THE FUTURE
OF WHOSE MARRIAGE?

Both Uncle Honoré and Gigi's grandmother remembered it well, according to Alan Jay Lerner's lyric. And this is what it had been like according to Uncle Honoré: "it was a lovely moonlit evening in May. You arrived at nine o'clock in your gold dress only a little late for our dinner engagement with friends. Afterwards there was that delightful carriage ride when we were so engrossed in one another that we didn't notice you had lost your glove." Ah, yes, Uncle Honoré remembered it well indeed, down to the last detail.

Or, come to think of it, did he? For Gigi's grandmother remembered it too, but not at all the same way. "There was no moon that rainy June evening. For once I was on time when we met at eight o'clock at the restaurant where we dined alone. You complimented me on my pretty blue dress. Afterwards we took a long walk and we were so engrossed in one another that we didn't notice I had lost my comb until my hair came tumbling down."

The Japanese motion picture *Rashomon* was built on the same idea—four different versions of

the same events. So, also, was Robert Gover's story of the college boy and the black prostitute in his *One Hundred Dollar Misunderstanding*. Also in this category is the old talmudic story of the learned rabbi called upon to render a decision in a marital situation. After listening carefully to the first spouse's story, he shook his head, saying, "You are absolutely right"; and, after listening equally carefully to the other spouse's story, he again shook his head, saying, "You are absolutely right."

There is no question in any of these examples of deliberate deceit or prevarication or insincerity or dishonesty. Both Uncle Honoré and Grandmamma are equally sincere, equally honest, equally "right." The discrepancies in their stories make a charming duet in *Gigi*. And even the happiest of mates can match such differences in their own memories.

In the case of Uncle Honoré and Grandmamma, we can explain the differences in the pictures they had in their heads of that evening half a century earlier: memories play strange tricks on all of us. But the same differences in the accounts of what happened show up also among modern couples even immediately after the event. In one study, for example, half of all the partners gave differing replies to questions about what had happened in a laboratory decision-making session they had just left. Other couples give different responses to questions about ordinary day-by-day events like lawn mowing as well as about romantic events. Once our attention has been called to the fact that both mates are equally sincere, equally honest, equally "right," the presence of two marriages in every marital union becomes clear—even obvious, as artists and wise persons have been telling us for so long.

Anyone, therefore, discussing the future of marriage has to specify whose marriage he is talking about: the husband's or the wife's. For there is by now a very considerable body of well-authenticated research to show that there really are two marriages in every marital union, and that they do not always coincide.

"HIS" AND "HER" MARRIAGES

Under the jargon "discrepant responses," the differences in the marriages of husbands and wives have come under the careful scrutiny of a score of researchers. They have found that when they ask husbands and wives identical questions about the union, they often get quite different replies. There is usually agreement on the number of children they have and a few other such verifiable items, although not, for example, on length of premarital acquaintance and of engagement, on age at marriage and interval between marriage and birth of first child. Indeed, with respect to even such basic components of the marriage as frequency of sexual relations, social interaction, household tasks, and decision making, they seem to be reporting on different marriages. As, I think, they are.

In the area of sexual relations, for example, Kinsey and his associates found different responses in from one- to two-thirds of the couples they studied. Kinsey interpreted these differences in terms of selective perception. In the generation he was studying, husbands wanted sexual relations

oftener than the wives did, thus "the females may be overestimating the actual frequencies" and "the husbands . . . are probably underestimating the frequencies." The differences might also have been vestiges of the probable situation earlier in the marriage when the desired frequency of sexual relations was about six to seven times greater among husbands than among wives. The difference may have become so impressed on the spouses that it remained in their minds even after the difference itself had disappeared or even been reversed. In a sample of happily married, middle-class couples a generation later, Harold Feldman found that both spouses attributed to their mates more influence in the area of sex than they did to themselves.

Companionship, as reflected in talking together, he found, was another area where differences showed up. Replies differed on three-fourths of all the items studied, including the topics talked about, the amount of time spent talking with each other, and which partner initiated conversation. Both partners claimed that whereas they talked more about topics of interest to their mates, their mates initiated conversations about topics primarily of interest to themselves. Harold Feldman concluded that projection in terms of needs was distorting even simple, everyday events, and lack of communication was permitting the distortions to continue. It seemed to him that "if these sex differences can occur so often among these generally well satisfied couples, it would not be surprising to find even less consensus and more distortion in other less satisfied couples."

Although, by and large, husbands and wives tend to become more alike with age, in this study of middle-class couples, differences increased with length of marriage rather than decreased, as one might logically have expected. More couples in the later than in the earlier years, for example, had differing pictures in their heads about how often they laughed together, discussed together, exchanged ideas, or worked together on projects, and about how well things were going between them.

The special nature of sex and the amorphousness of social interaction help to explain why differences in response might occur. But household tasks? They are fairly objective and clear-cut and not all that emotion-laden. Yet even here there are his-and-her versions. Since the division of labor in the household is becoming increasingly an issue in marriage, the uncovering of differing replies in this area is especially relevant. Hard as it is to believe, Granbois and Willett tell us that more than half of the partners in one sample disagreed on who kept track of money and bills. On the question, who mows the lawn? more than a fourth disagreed. Even family income was not universally agreed on.

These differences about sexual relations, companionship, and domestic duties tell us a great deal about the two marriages. But power or decision making can cover all aspects of a relationship. The question of who makes decisions or who exercises power has therefore attracted a great deal of research attention. If we were interested in who really had the power or who really made the decisions, the research would be hopeless. Would it be possible to draw any conclusion from a situation in which both partners agree that the husband ordered the wife to make all the decisions? Still, an enormous literature documents the quest of researchers for answers to the question of marital power. The major contribution it has made has been to reveal the existence of differences in replies between husbands and wives.

The presence of such inconsistent replies did not at first cause much concern. The researchers apologized for them but interpreted them as due to methodological inadequacies; if only they could find a better way to approach the problem, the differences would disappear. Alternatively, the use of only the wife's responses, which were more easily available, was justified on the grounds that differences in one direction between the partners in one marriage compensated for differences in another direction between the partners in another marriage and thus canceled them out. As, indeed,

they did. For when Granbois and Willett, two market researchers, analyzed the replies of husbands and wives separately, the overall picture was in fact the same for both wives and husbands. Such canceling out of differences in the total sample, however, concealed almost as much as it revealed about the individual couples who composed it. Granbois and Willett concluded, as Kinsey had earlier, that the "discrepancies . . . reflect differing perceptions on the part of responding partners." And this was the heart of the matter.

Differing reactions to common situations, it should be noted, are not at all uncommon. They are recognized in the folk wisdom embedded in the story of the blind men all giving different replies to questions on the nature of the elephant. One of the oldest experiments in juridical psychology demonstrates how different the statements of witnesses of the same act can be. Even in laboratory studies, it takes intensive training of raters to make it possible for them to arrive at agreement on the behavior they observe.

It has long been known that people with different backgrounds see things differently. We know, for example, that poor children perceive coins as larger than do children from more affluent homes. Boys and girls perceive differently. A good deal of the foundation for projective tests rests on the different ways in which individuals see identical stimuli. And this perception—or, as the sociologists put it, definition of the situation—is reality for them. In this sense, the realities of the husband's marriage are different from those of the wife's.

Finally, one of the most perceptive of the researchers, Constantina Safilios-Rothschild, asked the crucial question: Was what they were getting, even with the best research techniques, family sociology or wives' family sociology? She answered her own question: What the researchers who relied on wives' replies exclusively were reporting on was the wife's marriage. The husband's was not necessarily the same. There were, in fact, two marriages present:

One explanation of the discrepancies between the responses of husbands and wives may be the possibility of two "realities," the husband's subjective reality and the wife's subjective reality—two perspectives which do not always coincide. Each spouse perceives "facts" and situations differently according to his own needs, values, attitudes, and beliefs. An "objective" reality could possibly exist only in the trained observer's evaluation, if it does exist at all.

Interpreting the different replies of husbands and wives in terms of selective perception, projection of needs, values, attitudes, and beliefs, or different definitions of the situation, by no means renders them trivial or incidental or justifies dismissing or ignoring them. They are, rather, fundamental for an understanding of the two marriages, his and hers, and we ignore them at the peril of serious misunderstanding of marriage, present as well as future.

IS THERE AN OBJECTIVE REALITY IN MARRIAGE?

Whether or not husbands and wives perceive differently or define situations differently, still sexual relations are taking place, companionship is or is not occurring, tasks about the house are being performed, and decisions are being made every day by someone. In this sense, some sort of "reality" does exist. David Olson went to the laboratory to see if he could uncover it.

He first asked young couples expecting babies such questions as these: Which one of them would decide whether to buy insurance for the newborn child? Which one would decide the husband's part in diaper changing? Which one would decide whether the new mother would return to work or to school? When there were differences in the answers each gave individually on the questionnaire, he set up a situation in which together they had to arrive at a decision in

his laboratory. He could then compare the results of the questionnaire with the results in the simulated situation. He found neither spouse's questionnaire response any more accurate than the other's; that is, neither conformed better to the behavioral "reality" of the laboratory than the other did.

The most interesting thing, however, was that husbands, as shown on their questionnaire response, perceived themselves as having more power than they actually did have in the laboratory "reality," and wives perceived that they had less. Thus, whereas three-fourths (73 percent) of the husbands overestimated their power in decision making, 70 percent of the wives underestimated theirs. Turk and Bell found similar results in Canada. Both spouses tend to attribute decision-making power to the one who has the "right" to make the decision. Their replies, that is, conform to the model of marriage that has characterized civilized mankind for millennia. It is this model rather than their own actual behavior that husbands and wives tend to perceive.

We are now zeroing in on the basic reality. We can remove the quotation marks. For there is, in fact, an objective reality in marriage. It is a reality that resides in the cultural—legal, moral, and conventional—prescriptions and proscriptions and, hence, expectations that constitute marriage. It is the reality that is reflected in the minds of the spouses themselves. The differences between the marriages of husbands and of wives are structural realities, and it is these structural differences that constitute the basis for the different psychological realities.

THE AUTHORITY STRUCTURE OF MARRIAGE

Authority is an institutional phenomenon; it is strongly bound up with faith. It must be believed in; it cannot be enforced unless it also has power. Authority resides not in the person on whom it is conferred by the group or society, but in the recognition and acceptance it elicits in others. Power, on the other hand, may dispense with the prop of authority. It may take the form of the ability to coerce or to veto; it is often personal, charismatic, not institutional. This kind of personal power is self-enforcing. It does not require shoring up by access to force. In fact, it may even operate subversively. A woman with this kind of power may or may not know that she possesses it. If she does know she has it, she will probably disguise her exercise of it.

In the West, the institutional structure of marriage has invested the husband with authority and backed it by the power of church and state. The marriages of wives have thus been officially dominated by the husband. Hebrew, Christian, and Islamic versions of deity were in complete accord on this matter. The laws, written or unwritten, religious or civil, which have defined the marital union have been based on male conceptions, and they have undergirded male authority.

Adam came first. Eve was created to supply him with companionship, not vice versa. And God himself had told her that Adam would rule over her; her wishes had to conform to his. The New Testament authors agreed. Women were created for men, not men for women; women were therefore commanded to be obedient. If they wanted to learn anything, let them ask their husbands in private, for it was shameful for them to talk in the church. They should submit themselves to their husbands, because husbands were superior to wives; and wives should be as subject to their husbands as the church was to Christ. Timothy wrapped it all up: "Let the woman learn in silence with all subjection. But I suffer not a woman to teach, nor to usurp authority over the man, but to be in silence." Male Jews continued for millennia to thank God three times a day that they were not women. And the Koran teaches women that men are naturally their superiors because God made them that way; naturally, their own status is one of subordination.

The state as well as the church had the same conception of marriage, assigning to the husband and father control over his dependents, including his wife. Sometimes this power was well-nigh absolute, as in the case of the Roman patria potestas—or the English common law, which flatly said, "The husband and wife are as one and that one is the husband." There are rules still lingering today with the same, though less extreme, slant. Diane B. Schulder has summarized the legal framework of the wife's marriage as laid down in the common law.

> The legal responsibilities of a wife are to live in the home established by her husband; to perform the domestic chores (cleaning, cooking, washing, etc.) necessary to help maintain that home; to care for her husband and children. . . . A husband may force his wife to have sexual relations as long as his demands are reasonable and her health is not endangered. . . . The law allows a wife to take a job if she wishes. However, she must see that her domestic chores are completed, and if there are children, that they receive proper care during her absence.

A wife is not entitled to payment for household work; and some jurisdictions in the United States expressly deny payment for it. In some states, the wife's earnings are under the control of her husband, and in four, special court approval and in some cases husband's consent are required if a wife wishes to start a business of her own.

The male counterpart to these obligations includes that of supporting his wife. He may not disinherit her. She has a third interest in property owned by him, even if it is held in his name only. Her name is required when he sells property.

Not only divine and civil law but also rules of etiquette have defined authority as a husband's prerogative. One of the first books published in English was a *Boke of Good Manners*, translated from the French of Jacques Le Grand in 1487, which included a chapter on "How Wymmen Ought to Be Gouerned." The thirty-third rule of Plutarch's *Rules for Husbands and Wives* was that women should obey their husbands; if they "try to rule over their husbands they make a worse mistake than the husbands do who let themselves be ruled." The husband's rule should not, of course, be brutal; he should not rule his wife "as a master does his chattel, but as the soul governs the body, by feeling with her and being linked to her by affection." Wives, according to Richard Baxter, a seventeenth-century English divine, had to obey even a wicked husband, the only exception being that a wife need not obey a husband if he ordered her to change her religion. But, again, like Plutarch, Baxter warned that the husband should love his wife; his authority should not be so coercive or so harsh as to destroy love. Among his twelve rules for carrying out the duties of conjugal love, however, was one to the effect that love must not be so imprudent as to destroy authority.

As late as the nineteenth century, Tocqueville noted that in the United States the ideals of democracy did not apply between husbands and wives:

> Nor have the Americans ever supposed that one consequence of democratic principles is the subversion of marital power, or the confusion of the natural authorities in families. They hold that every association must have a head in order to accomplish its objective, and that the natural head of the conjugal association is man. They do not therefore deny him the right of directing his partner; and they maintain, that in the smaller association of husband and wife, as well as in the great social community, the object of democracy is to regulate and legalize the powers which are necessary, not to subvert all power.
>
> This opinion is not peculiar to men and contested by women; I never observed that the women of America consider conjugal authority as a fortunate usurpation [by men] of their rights, nor that they thought themselves degraded by submitting to it. It appears to me, on the contrary, that they attach a sort of pride to the voluntary surrender of their own will and make it their

boast to bend themselves to the yoke, not to shake it off.

The point here is not to document once more the specific ways (religious, legal, moral, traditional) in which male authority has been built into the marital union—that has been done a great many times—but merely to illustrate how different (structurally or "objectively" as well as perceptually or "subjectively") the wife's marriage has actually been from the husband's throughout history.

THE SUBVERSIVENESS
OF NATURE

The rationale for male authority rested not only on biblical grounds but also on nature or natural law, on the generally accepted natural superiority of men. For nothing could be more self-evident than that the patriarchal conception of marriage, in which the husband was unequivocally the boss, was natural, resting as it did on the unchallenged superiority of males.

Actually, nature, if not deity, is subversive. Power, or the ability to coerce or to veto, is widely distributed in both sexes, among women as well as among men. And whatever the theoretical or conceptual picture may have been, the actual, day-by-day relationships between husbands and wives have been determined by the men and women themselves. All that the institutional machinery could do was to confer authority; it could not create personal power, for such power cannot be conferred, and women can generate it as well as men. Thus, keeping women in their place has been a universal problem, in spite of the fact that almost without exception institutional patterns give men positions of superiority over them.

If the sexes were, in fact, categorically distinct, with no overlapping, so that no man was inferior to any woman or any woman superior to any man, or vice versa, marriage would have been a great deal simpler. But there is no such sharp cleavage between the sexes except with respect to the presence or absence of certain organs. With all the other characteristics of each sex, there is greater or less overlapping, some men being more "feminine" than the average woman and some women more "masculine" than the average man. The structure of families and societies reflects the positions assigned to men and women. The bottom stratum includes children, slaves, servants, and outcasts of all kinds, males as well as females. As one ascends the structural hierarchy, the proportion of males increases, so that at the apex there are only males.

When societies fall back on the lazy expedient—as all societies everywhere have done—of allocating the rewards and punishments of life on the basis of sex, they are bound to create a host of anomalies, square pegs in round holes, societal misfits. Roles have been allocated on the basis of sex which did not fit a sizable number of both sexes—women, for example, who chafed at subordinate status and men who could not master superordinate status. The history of the relations of the sexes is replete with examples of such misfits. Unless a *modus vivendi* is arrived at, unhappy marriages are the result.

There is, though, a difference between the exercise of power by husbands and by wives. When women exert power, they are not rewarded; they may even be punished. They are "deviant." Turk and Bell note that "wives who . . . have the greater influence in decision making may experience guilt over this fact." They must therefore dissemble to maintain the illusion, even to themselves, that they are subservient. They tend to feel less powerful than they are because they *ought* to be.

When men exert power, on the other hand, they are rewarded; it is the natural expression of authority. They feel no guilt about it. The prestige of authority goes to the husband whether or not he is actually the one who exercises it. It is not often even noticed when the wife does so. She sees to it that it is not.

There are two marriages, then, in every marital union, his and hers. . . . The questions, therefore, are these: In what direction will they change in the future? Will one change more than the other? Will they tend to converge or to diverge? Will the future continue to favor the husband's marriage? And if the wife's marriage is improved, will it cost the husband's anything, or will his benefit along with hers?

DISCUSSION QUESTIONS

1. Bernard describes marriage as a social institution that changes over time—both within given marriage relationships and as a social institution. What changes in the institution of marriage have occurred in the time since Bernard wrote her essay? Do men and women still have different experiences within marriage? If so, what are they and why?

2. Bernard describes marriage, in part, as a power relationship. How does gender influence the power of men and women within marriage? Do similar dynamics occur in other forms of intimate relationships?

3. Bernard describes marriage as a *gendered institution,* meaning that as a social form it embeds characteristics associated with the social meaning of gender in society. How do the social constructs of "man" and "woman" influence the nature of marriage relationships in society?

INTERNET RESOURCES

Suggested Web URLs
for Further Study

*http://www.census.gov/population/www/socdemo
/ms-la.html*
US Census Marital Status and Living Arrangements. This site from the US Census Bureau provides an excellent picture of current marital status and living arrangements in the United States.

http://www.trinity.edu/mkearl/gender.html
Gender and Society. This is a very user-friendly site for students that features web links integrated into the text of a summary presentation about gender.

http://www.trinity.edu/mkearl/family.html
An academic sociology site created by Michael C. Kearl at Trinity University in San Antonio, this is a comprehensive resource for the study of marriage and family processes.

http://www.aamft.org
American Association for Marriage and Family Therapy. This professional association of marriage and family therapists provides resources for therapists and clients, as well as information about families and health.

InfoTrac College Edition

You can find further relevant readings on the World Wide Web at
http://sociology.wadsworth.com

Virtual Society

For further information on this subject including links to relevant Web sites, go to the Wadsworth Sociology homepage at
http://sociology.wadsworth.com

12

On Being Sane in Insane Places

D. L. ROSENHAN

If sanity and insanity exist, how shall we know them?

The question is neither capricious nor itself insane. However much we may be personally convinced that we can tell the normal from the abnormal, the evidence is simply not compelling. It is commonplace, for example, to read about murder trials wherein eminent psychiatrists for the defense are contradicted by equally eminent psychiatrists for the prosecution on the matter of the defendant's sanity. More generally, there are a great deal of conflicting data on the reliability, utility, and meaning of such terms as "sanity," "insanity," "mental illness," and "schizophrenia." Finally, as early as 1934, Benedict suggested that normality and abnormality are not universal. What is viewed as normal in one culture may be seen as quite aberrant in another. Thus, notions of normality and abnormality may not be quite as accurate as people believe they are.

To raise questions regarding normality and abnormality is in no way to question the fact that some behaviors are deviant or odd. Murder is deviant. So, too, are hallucinations. Nor does raising such questions deny the existence of the personal anguish that is often associated with "mental illness." Anxiety and depression exist. Psychological suffering exists. But normality and abnormality, sanity and insanity, and the diagnoses that flow from them may be less substantive than many believe them to be.

At its heart, the question of whether the sane can be distinguished from the insane (and whether degrees of insanity can be distinguished from each other) is a simple matter: do the salient characteristics that lead to diagnoses reside in the patients themselves or in the environments and contexts in which observers find them? From Bleuler, through Kretchmer, through the formulators of the recently revised *Diagnostic and Statistical Manual* of the American Psychiatric Association, the belief has been strong that patients present symptoms, that those symptoms can be categorized, and, implicitly, that the sane are distinguishable from the insane. More recently, however, this belief has been questioned. Based in part on theoretical and anthropological considerations, but also on philosophical, legal, and therapeutic ones, the view has grown that psychological categorization of mental illness is useless at best and downright harmful, misleading, and pejorative at worst. Psychiatric diagnoses, in this view, are in the minds of the observers and are not valid summaries of characteristics displayed by the observed.

Gains can be made in deciding which of these is more nearly accurate by getting normal people (that is, people who do not have, and have never suffered, symptoms of serious psychiatric disorders) admitted to psychiatric hospitals and then determining whether they were discovered to be sane and, if so, how. If the sanity of such pseudopatients were always detected, there would be prima facie evidence that a sane individual can be distinguished from the insane context in which he is found. Normality (and presumably abnormality) is distinct enough that it can be recognized wherever it occurs, for it is carried within the person. If, on the other hand, the sanity of the pseudopatients were never discovered, serious difficulties would arise for those who support traditional modes of psychiatric diagnosis. Given that the hospital staff was not incompetent, that the pseudopatient had been behaving as sanely as he had been outside of the hospital, and that it had never been previously suggested that he belonged in a psychiatric hospital, such an unlikely outcome would support the view that psychiatric diagnosis betrays little about the patient but much about the environment in which an observer finds him.

This article describes such an experiment. Eight sane people gained secret admission to 12 different hospitals. Their diagnostic experiences constitute the data of the first part of this article; the remainder is devoted to a description of their experiences in psychiatric institutions. Too few psychiatrists and psychologists, even those who have worked in such hospitals, know what the experience is like. They rarely talk about it with former patients, perhaps because they distrust information coming from the previously insane. Those who have worked in psychiatric hospitals are likely to have adapted so thoroughly to the settings that they are insensitive to the impact of that experience. And while there have been occasional reports of researchers who submitted themselves to psychiatric hospitalization, these researchers have commonly remained in the hospitals for short periods of time, often with the knowledge of the hospital staff. It is difficult to know the extent to which they were treated like

patients or like research colleagues. Nevertheless, their reports about the inside of the psychiatric hospital have been valuable. This article extends those efforts.

PSEUDOPATIENTS AND THEIR SETTINGS

The eight pseudopatients were a varied group. One was a psychology graduate student in his 20s. The remaining seven were older and "established." Among them were three psychologists, a pediatrician, a psychiatrist, a painter, and a housewife. Three pseudopatients were women, five were men. All of them employed pseudonyms, lest their alleged diagnoses embarrass them later. Those who were in mental health professions alleged another occupation in order to avoid the special attentions that might be accorded by staff, as a matter of courtesy or caution, to ailing colleagues. With the exception of myself (I was the first pseudopatient and my presence was known to the hospital administrator and chief psychologist and, so far as I can tell, to them alone), the presence of pseudopatients and the nature of the research program was not known to the hospital staffs.

The settings were similarly varied. In order to generalize the findings, admission into a variety of hospitals was sought. The 12 hospitals in the sample were located in five different states on the East and West coasts. Some were old and shabby, some were quite new. Some were research-oriented, others not. Some had good staff-patient ratios, others were quite understaffed. Only one was a strictly private hospital. All of the others were supported by state or federal funds or, in one instance, by university funds.

After calling the hospital for an appointment, the pseudopatient arrived at the admissions office complaining that he had been hearing voices. Asked what the voices said, he replied that they were often unclear, but as far as he could tell they said "empty," "hollow," and "thud." The voices were unfamiliar and were of the same sex as the pseudopatient. The choice of these symptoms was occasioned by their apparent similarity to existential symptoms. Such symptoms are alleged to arise from painful concerns about the perceived meaninglessness of one's life. It is as if the hallucinating person were saying, "My life is empty and hollow." The choice of these symptoms was also determined by the *absence* of a single report of existential psychoses in the literature.

Beyond alleging the symptoms and falsifying name, vocation, and employment, no further alterations of person, history, or circumstances were made. The significant events of the pseudopatient's life history were presented as they had actually occurred. Relationships with parents and siblings, with spouse and children, with people at work and in school, consistent with the aforementioned exceptions, were described as they were or had been. Frustrations and upsets were described along with joys and satisfactions. These facts are important to remember. If anything, they strongly biased the subsequent results in favor of detecting sanity, since none of their histories or current behaviors were seriously pathological in any way.

Immediately upon admission to the psychiatric ward, the pseudopatient ceased simulating *any* symptoms of abnormality. In some cases, there was a brief period of mild nervousness and anxiety, since none of the pseudopatients really believed that they would be admitted so easily. Indeed, their shared fear was that they would be immediately exposed as frauds and greatly embarrassed. Moreover, many of them had never visited a psychiatric ward; even those who had, nevertheless had some genuine fears about what might happen to them. Their nervousness, then, was quite appropriate to the novelty of the hospital setting, and it abated rapidly.

Apart from that short-lived nervousness, the pseudopatient behaved on the ward as he "normally" behaved. The pseudopatient spoke to patients and staff as he might ordinarily. Because there is uncommonly little to do on a psychiatric ward, he attempted to engage others in conversation. When asked by staff how he was feeling, he indicated that he was fine, that he no longer

experienced symptoms. He responded to instructions from attendants, to calls for medication (which was not swallowed), and to dining-hall instructions. Beyond such activities as were available to him on the admissions ward, he spent his time writing down his observations about the ward, its patients, and the staff. Initially these notes were written "secretly," but as it soon became clear that no one much cared, they were subsequently written on standard tablets of paper in such public places as the dayroom. No secret was made of these activities.

The pseudopatient, very much as a true psychiatric patient, entered a hospital with no foreknowledge of when he would be discharged. Each was told that he would have to get out by his own devices, essentially by convincing the staff that he was sane. The psychological stresses associated with hospitalization were considerable, and all but one of the pseudopatients desired to be discharged almost immediately after being admitted. They were, therefore, motivated not only to behave sanely, but to be paragons of cooperation. That their behavior was in no way disruptive is confirmed by nursing reports, which have been obtained on most of the patients. These reports uniformly indicate that the patients were "friendly," "cooperative," and "exhibited no abnormal indications."

THE NORMAL ARE NOT DETECTABLY SANE

Despite their public "show" of sanity, the pseudopatients were never detected. Admitted, except in one case, with a diagnosis of schizophrenia, each was discharged with a diagnosis of schizophrenia "in remission." The label "in remission" should in no way be dismissed as a formality, for at no time during any hospitalization had any question been raised about any pseudopatient's simulation. Nor are there any indications in the hospital records that the pseudopatient's status was suspect. Rather, the evidence is strong that, once labeled schizophrenic, the pseudo-patient was stuck with that label. If the pseudo-patient was to be discharged, he must naturally be "in remission"; but he was not sane, nor, in the institution's view, had he ever been sane.

The uniform failure of recognize sanity cannot be attributed to the quality of the hospitals, for although there were considerable variations among them, several are considered excellent. Nor can it be alleged that there was simply not enough time to observe the pseudopatients. Length of hospitalization ranged from 7 to 52 days, with an average of 19 days. The pseudo-patients were not, in fact, carefully observed, but this failure clearly speaks more to traditions within psychiatric hospitals than to lack of opportunity.

Finally, it cannot be said that the failure to recognize the pseudopatients' sanity was due to the fact that they were not behaving sanely. While there was clearly some tension present in all of them, their daily visitors could detect no serious behavioral consequences—nor, indeed, could other patients. It was quite common for the patients to "detect" the pseudopatients' sanity. During the first three hospitalizations, when accurate counts were kept, 35 of a total of 118 patients on the admissions ward voiced their suspicions, some vigorously, "You're not crazy. You're a journalist, or a professor [referring to the continual note-taking]. You're checking up on the hospital." While most of the patients were reassured by the pseudopatient's insistence that he had been sick before he came in but was fine now, some continued to believe that the pseudopatient was sane throughout his hospitalization. The fact that the patients often recognized normality when staff did not raises important questions.

Failure to detect sanity during the course of the hospitalization may be due to the fact that physicians operate with a strong bias toward what statisticians call the type 2 error. This is to say that physicians are more inclined to call a healthy person sick (a false positive, type 2) than a sick person healthy (a false negative, type 1). The reasons for this are not hard to find: it is clearly more dangerous to misdiagnose illness than health. Better to err on the side of caution, to suspect illness even among the healthy.

But what holds for medicine does not hold equally well for psychiatry. Medical illnesses, while unfortunate, are not commonly pejorative. Psychiatric diagnoses, on the contrary, carry with them personal, legal, and social stigmas. It was therefore important to see whether the tendency toward diagnosing the sane insane could be reversed. The following experiment was arranged at a research and teaching hospital whose staff had heard these findings but doubted that such an error could occur in their hospital. The staff was informed that at some time during the following 3 months, one or more pseudopatients would attempt to be admitted into the psychiatric hospital. Each staff member was asked to rate each patient who presented himself at admissions or on the ward according to the likelihood that the patient was a pseudopatient. A 10-point scale was used, with a 1 and 2 reflecting high confidence that the patient was a pseudopatient.

Judgments were obtained on 193 patients who were admitted for psychiatric treatment. All staff who had had sustained contact with or primary responsibility for the patient—attendants, nurses, psychiatrists, physicians, and psychologists—were asked to make judgments. Forty-one patients were alleged, with high confidence, to be pseudopatients by at least one member of the staff. Twenty-three were considered suspect by at least one psychiatrist. Nineteen were suspected by one psychiatrist *and* one other staff member. Actually, no genuine pseudopatient (at least from my group) presented himself during this period.

The experiment is instructive. It indicates that the tendency to designate sane people as insane can be reversed when the stakes (in this case, prestige and diagnostic acumen) are high. But what can be said of the 19 people who were suspected of being "sane" by one psychiatrist and another staff member? Were these people truly "sane," or was it rather the case that in the course of avoiding the type 2 error of the staff tended to make more errors of the first sort—calling the crazy "sane"? There is no way of knowing. But one thing is certain: any diagnostic process that lends itself so readily to massive errors of this sort cannot be a very reliable one.

THE STICKINESS OF PSYCHODIAGNOSTIC LABELS

Beyond the tendency to call the healthy sick—a tendency that accounts better for diagnostic behavior on admission than it does for such behavior after a lengthy period of exposure—the data speak to the massive role of labeling in psychiatric assessment. Having once been labeled schizophrenic, there is nothing the pseudopatient can do to overcome the tag. The tag profoundly colors others' perceptions of him and his behavior. . . .

Once a person is designated abnormal, all of his other behaviors and characteristics are colored by that label. Indeed, that label is so powerful that many of the pseudopatients' normal behaviors were overlooked entirely or profoundly misinterpreted. Some examples may clarify this issue.

Earlier I indicated that there were no changes in the pseudopatient's personal history and current status beyond those of name, employment, and, where necessary, vocation. Otherwise, a veridical description of personal history and circumstances was offered. Those circumstances were not psychotic. How were they made consonant with the diagnosis of psychosis? Or were those diagnoses modified in such a way as to bring them into accord with the circumstances of the pseudopatient's life, as described by him?

As far as I can determine, diagnoses were in no way affected by the relative health of the circumstances of a pseudopatient's life. Rather, the reverse occurred: the perception of his circumstances was shaped entirely by the diagnosis. A clear example of such translation is found in the case of a pseudopatient who had had a close relationship with his mother but was rather remote from his father during his early childhood. During adolescence and beyond, however, his father became a close friend, while his relationship with his mother cooled. His present relationship with his wife was characteristically close and warm. Apart from occasional angry exchanges, friction was minimal. The children had rarely been spanked. Surely there is nothing especially pathological about such a history. Indeed, many readers may see a similar pattern in their own experiences, with no markedly deleterious consequences. Observe, however, how such a history was translated in the psychopathological context, this from the case summary prepared after the patient was discharged.

> This white 39-year-old male . . . manifests a long history of considerable ambivalence in close relationships, which begins in early childhood. A warm relationship with his mother cools during his adolescence. A distant relationship to his father is described as becoming very intense. Affective stability is absent. His attempts to control emotionality with his wife and children are punctuated by angry outbursts and, in the case of the children, spankings. And while he says that he has several good friends, one senses considerable ambivalence embedded in those relationships also. . . .

The facts of the case were unintentionally distorted by the staff to achieve consistency with a popular theory of the dynamics of a schizophrenic reaction. Nothing of an ambivalent nature has been described in relations with parents, spouse, or friends. To the extent that ambivalence could be inferred, it was probably not greater than is found in all human relationships. It is true the pseudopatient's relationships with his parents changed over time, but in the ordinary context that would hardly be remarkable—indeed, it might very well be expected. Clearly, the meaning ascribed to his verbalizations (that is, ambivalence, affective instability) was determined by the diagnosis: schizophrenia. An entirely different meaning would have been ascribed if it were known that the man was "normal."

All pseudopatients took extensive notes publicly. Under ordinary circumstances, such behavior would have raised questions in the minds of observers, as, in fact, it did among patients. Indeed, it seemed so certain that the notes would elicit suspicion that elaborate precautions were taken to remove them from the ward each day. But the precautions proved needless. The closest any staff member came to questioning these notes occurred when one pseudopatient asked his physician what kind of medication he was receiving and began to write down the response. "You needn't write it," he was told gently. "If you have trouble remembering, just ask me again."

If no questions were asked of the pseudopatients, how was their writing interpreted? Nursing records for three patients indicate that the writing was seen as an aspect of their pathological behavior. "Patient engages in writing behavior" was the daily nursing comment on one of the pseudopatients who was never questioned about his writing. Given that the patient is in the hospital, he must be psychologically disturbed. And given that he is disturbed, continuous writing must be a behavioral manifestation of that disturbance, perhaps a subset of the compulsive behaviors that are sometimes correlated with schizophrenia.

One tacit characteristic of the psychiatric diagnosis is that it locates the sources of aberration within the individual and only rarely within the complex of stimuli that surrounds him. Consequently, behaviors that are stimulated by the environment are commonly misattributed to the patient's disorder. For example, one kindly nurse found a pseudopatient pacing the long hospital corridors. "Nervous, Mr. X?" she asked. "No, bored," he said.

The notes kept by pseudopatients are full of patient behaviors that were misinterpreted by well-intentioned staff. Often enough, a patient would go "berserk" because he had, wittingly or unwittingly, been mistreated by, say, an attendant. A nurse coming upon the scene would rarely inquire even cursorily into the environmental stimuli of the patient's behavior. Rather, she assumed that his upset derived from his pathology, not from his present interactions with other staff members. Occasionally, the staff might assume that the patient's family (especially when they had recently visited) or other patients had stimulated the outburst. But never were the staff found to assume that one of themselves or the structure of the hospital had anything to do with a patient's behavior. One psychiatrist pointed to a group of patients who were sitting outside the cafeteria entrance half an hour before lunchtime. To a group of young residents he indicated that such behavior was characteristic of the oral-acquisitive nature of the syndrome. It seemed not to occur to him that there were very few things to anticipate in a psychiatric hospital besides eating.

A psychiatric label has a life and an influence of its own. Once the impression has been formed that the patient is schizophrenic, the expectation is that he will continue to be schizophrenic. When a sufficient amount of time has passed, during which the patient has done nothing bizarre, he is considered to be in remission and available for discharge. But the label endures beyond discharge, with the unconfirmed expectation that he will behave as a schizophrenic again. Such labels, conferred by mental health professionals, are as influential on the patient as they are on his relatives and friends, and it should not surprise anyone that the diagnosis acts on all of them as a self-fulfilling prophecy. Eventually, the patient himself accepts the diagnosis with all of its surplus meanings and expectations, and behaves accordingly.

The inferences to be made from these matters are quite simple. Much as Zigler and Phillips have demonstrated that there is enormous overlap in the symptoms presented by patients who have been variously diagnosed, so there is enormous overlap in the behaviors of the sane and the insane. The sane are not "sane" all of the time. We lose our tempers "for no good reason." We are occasionally depressed or anxious, again for no good reason. And we may find it difficult to get along with one or another person—again for no reason that we can specify. Similarly, the insane are not always insane. Indeed, it was the impression of the pseudopatients while living with them that they were sane for long periods of time—that the bizarre behaviors upon which their diagnoses were allegedly predicated constituted only a small fraction of their total behavior. If it makes no sense to label ourselves permanently depressed on the basis of an occasional depression, then it takes better evidence than is presently available to label all patients insane or schizophrenic on the basis of bizarre behaviors or cognitions. It seems more useful, as Mischel has pointed out, to limit our discussions to *behavior,* the stimuli that provoke them, and their correlates.

It is not known why powerful impressions of personality traits, such as "crazy" or "insane," arise. Conceivably, when the origins of and stimuli that give rise to a behavior are remote or unknown, or when the behavior strikes us as immutable. trait labels regarding the *behaver* arise. When, on the other hand, the origins and stimuli are known and available, discourse is limited to the behavior itself. Thus, I may hallucinate because I am sleeping, or I may hallucinate because I have ingested a peculiar drug. These are termed sleep-induced hallucinations, or dreams, and drug-induced hallucinations, respectively. But when the stimuli to my hallucinations are unknown, that is called craziness, or schizophrenia—as if that inference were somehow as illuminating as the others.

★ ★ ★

THE CONSEQUENCES
OF LABELING . . .

Whenever the ratio of what is known to what needs to be known approaches zero, we tend to invent "knowledge" and assume that we understand more than we actually do. We seem unable to acknowledge that we simply don't know. The needs for diagnosis and remediation of behavioral and emotional problems are enormous. But

rather than acknowledge that we are just embarking on understanding, we continue to label patients "schizophrenic," "manic-depressive," and "insane," as if in those words we had captured the essence of understanding. The facts of the matter are that we have known for a long time that diagnoses are often not useful or reliable, but we have nevertheless continued to use them. We now know that we cannot distinguish insanity from sanity. It is depressing to consider how that information will be used.

Not merely depressing, but frightening. How many people, one wonders, are sane but not recognized as such in our psychiatric institutions? How many have been needlessly stripped of their privileges of citizenship, from the right to vote and drive to that of handling their own accounts. How many have feigned insanity in order to avoid the criminal consequences of their behavior, and, conversely, how many would rather stand trial than live interminably in a psychiatric hospital—but are wrongly thought to be mentally ill? How many have been stigmatized by well-intentioned, but nevertheless erroneous, diagnoses? One the last point, recall again that a "type 2 error" in psychiatric diagnosis does not have the same consequences it does in medical diagnosis. A diagnosis of cancer that has been found to be in error is cause for celebration. But psychiatric diagnoses are rarely found to be in error. The label sticks, a mark of inadequacy forever.

Finally, how many patients might be "sane" outside the psychiatric hospital but seem insane in it—not because craziness resides in them, as it were, but because they are responding to a bizarre setting, one that may be unique to institutions which harbor nether people? Goffman calls the process of socialization to such institutions "mortification". . . . And while it is impossible to known whether the pseudopatients' responses to these processes are characteristic of all inmates—they were, after all, not real patients—it is difficult to believe that these processes of socialization to a psychiatric hospital provide useful attitudes or habits of response for living in the "real world."

SUMMARY
AND CONCLUSIONS

It is clear that we cannot distinguish the sane from the insane in psychiatric hospitals. The hospital itself imposes a special environment in which the meanings of behavior can easily be misunderstood. The consequences to patients hospitalized in such an environment—the powerlessness, depersonalization, segregation, mortification, and self-labeling—seem undoubtedly counter-therapeutic.

I do not, even now, understand this problem well enough to perceive solutions. But two matters seem to have some promise. The first concerns the proliferation of community mental health facilities, of crisis intervention centers, of the human potential movement, and of behavior therapies that, for all of their own problems, tend to avoid psychiatric labels, to focus on specific problems and behaviors, and to retain the individual in a relatively nonpejorative environment. Clearly, to the extent that we refrain from sending the distressed to insane places, our impressions of them are less likely to be distorted. (The risk of distorted perceptions, it seems to me, is always present, since we are much more sensitive to an individual's behaviors and verbalizations than we are to the subtle contextual stimuli that often promote them. At issue here is a matter of magnitude. And, as I have shown, the magnitude of distortion is exceedingly high in the extreme context that is a psychiatric hospital.)

The second matter that might prove promising speaks to the need to increase the sensitivity of mental health workers and researchers to the *Catch 22* position of psychiatric patients. Simply reading materials in this area will be of help to some such workers and researchers. For others, directly experiencing the impact of psychiatric hospitalization will be of enormous use. Clearly, further research into the social psychology of such total institutions will both facilitate treatment and deepen understanding.

I and the other pseudopatients in the psychiatric setting had distinctly negative reactions. We do not pretend to describe the subjective experiences of true patients. Theirs may be different from ours, particularly with the passage of time and the necessary process of adaptation to one's environment. But we can and do speak to the relatively more objective indices of treatment within the hospital. It could be a mistake, and a very unfortunate one, to consider that what happened to us derived from malice or stupidity on the part of the staff. Quite the contrary, our overwhelming impression of them was of people who really cared, who were committed and who were uncommonly intelligent. Where they failed, as they sometimes did painfully, it would be more accurate to attribute those failures to the environment in which they, too, found themselves than to personal callousness. Their perceptions and behavior were controlled by the situation, rather than being motivated by a malicious disposition. In a more benign environment, one that was less attached to global diagnosis, their behaviors and judgments might have been more benign and effective.

DISCUSSION QUESTIONS

1. Rosenhan's experiment indicates the importance of labeling the social construction of reality. That is, what is perceived as "real" is the result of people's reactions to certain behaviors. How does this explain other situations in which labeling is an ongoing part of social interaction?

2. In what kinds of environments other than hospitals and asylums might labeling take on the significance that it has in Rosenhan's study? What are the consequences?

3. Goffman's work on the presentation of self (see prior reading) suggests that people can

manipulate how they appear to others, but Rosenhan reminds us that there is always a risk of distorted perceptions by others. How might this explain how people are stereotyped by different social factors?

INTERNET RESOURCES

Suggested Web URLs
for Further Study

http://www.stigma.org/everyfamily/vyabolger.html
This Web site explores models of psychopathology and their relationship to stigmatization.

http://www.homeoint.org/morrell/otherarticles/marginality.htm
This site contains a presentation given at a conference on sociology on the subject of deviance, marginality and social exclusion.

http://www.awa.com/norton/struc/chap_18/chap_18.html
This is a Web site on psychopathology with links to various forms of mental illness and disorders.

InfoTrac College Edition

You can find further relevant readings on the World Wide Web at
http://sociology.wadsworth.com

Virtual Society

For further information on this subject including links to relevant Web sites, go to the Wadsworth Sociology homepage at
http://sociology.wadsworth.com

13

The Power Elite

C. WRIGHT MILLS

The powers of ordinary men are circumscribed by the everyday worlds in which they live, yet even in these rounds of job, family, and neighborhood they often seem driven by forces they can neither understand nor govern. "Great changes" are beyond their control, but affect their conduct and outlook nonetheless. The very framework of modern society confines them to projects not their own, but from every side, such changes now press upon the men and women of the mass society, who accordingly feel that they are without purpose in an epoch in which they are without power.

But not all men are in this sense ordinary. As the means of information and of power are centralized, some men come to occupy positions in American society from which they can look down upon, so to speak, and by their decisions mightily affect, the everyday worlds of ordinary men and women. They are not made by their jobs; they set up and break down jobs for thousands of others; they are not confined by simple family responsibilities; they can escape. They may live in many hotels and houses, but they are bound by no one community. They need not merely "meet the demands of the day and hour"; in some part, they create these demands, and cause others to meet them. Whether or not they profess their power, their technical and political experience of it far transcends that of the underlying population. What Jacob Burckhardt said of "great men," most Americans might well say of their elite: "They are all that we are not."

The power elite is composed of men whose positions enable them transcend the ordinary environments of ordinary men and women; they are in positions to make decisions having major consequences. Whether they do or do not make such decisions is less important than the fact that they do occupy such pivotal positions: Their failure to act, their failure to make decisions, is itself an act that is often of greater consequence than the decisions they do make. For they are in command of the major hierarchies and organizations of modern society. They rule the big corporations. They run the machinery of the state and claim its prerogatives. They direct the military establishment. They occupy the strategic command posts of the social structure, in which are now centered the effective means of the power and the wealth and the celebrity which they enjoy.

Credit: From *The Power Elite, New Edition* by C. Wright Mills, © 1956, 2000 by Oxford University Press, Inc. Used by permission of Oxford University Press, Inc.

The power elite are not solitary rulers. Advisers and consultants, spokesmen and opinion-makers are often the captains of their higher thought and decision. Immediately below the elite are the professional politicians of the middle levels of power, in the Congress and in the pressure groups, as well as among the new and old upper classes of town and city and region. Mingling with them, in curious ways which we shall explore, are those professional celebrities who live by being continually displayed but are never, so long as they remain celebrities, displayed enough. If such celebrities are not at the head of any dominating hierarchy, they do often have the power to distract the attention of the public or afford sensations to the masses, or, more directly, to gain the ear of those who do occupy positions of direct power. More or less unattached, as critics of morality and technicians of power, as spokesmen of God and creators of mass sensibility, such celebrities and consultants are part of the immediate scene in which the drama of the elite is enacted. But that drama itself is centered in the command posts of the major institutional hierarchies.

The truth about the nature and the power of the elite is not some secret which men of affairs know but will not tell. Such men hold quite various theories about their own roles in the sequence of event and decision. Often they are uncertain about their roles, and even more often they allow their fears and their hopes to affect their assessment of their own power. No matter how great their actual power, they tend to be less acutely aware of it than of the resistances of others to its use. Moreover, most American men of affairs have learned well the rhetoric of public relations, in some cases even to the point of using it when they are alone, and thus coming to believe it. The personal awareness of the actors is only one of the several sources one must examine in order to understand the higher circles. Yet many who believe that there is no elite, or at any rate none of any consequence, rest their argument upon what men of affairs believe about themselves, or at least assert in public.

There is, however, another view: Those who feel, even if vaguely, that a compact and powerful elite of great importance does now prevail in America often base that feeling upon the historical trend of our time. They have felt, for example, the domination of the military event, and from this they infer that generals and admirals, as well as other men of decision influenced by them, must be enormously powerful. They hear that the Congress has again abdicated to a handful of men decisions clearly related to the issue of war or peace. They know that the bomb was dropped over Japan in the name of the United States of America, although they were at no time consulted about the matter. They feel that they live in a time of big decisions; they know that they are not making any. Accordingly, as they consider the present as history, they infer that at its center, making decisions or failing to make them, there must be an elite of power.

On the one hand, those who share this feeling about big historical events assume that there is an elite and that its power is great. On the other hand, those who listen carefully to the reports of men apparently involved in the great decisions often do not believe that there is an elite whose powers are of decisive consequence.

Both views must be taken into account, but neither is adequate. The way to understand the power of the American elite lies neither solely in recognizing the historic scale of events nor in accepting the personal awareness reported by men of apparent decision. Behind such men and behind the events of history, linking the two, are the major institutions of modern society. These hierarchies of state and corporation and army constitute the means of power; as such they are now of a consequence not before equaled in human history—and at their summits, there are now those command posts of modern society which offer us the sociological key to an understanding of the role of the higher circles in America.

Within American society, major national power now resides in the economic, the political, and the military domains. Other institutions seem off to the side of modern history, and, on occasion, duly sub-ordinated to these. No family is as directly powerful in national affairs as any major corporation; no church is as directly powerful in the external biographies of young men in America today as the military establishment; no college is as powerful in the shaping of momentous events as the National Security Council. Religious, educational, and family institutions are not autonomous centers of national power; on the contrary, these decentralized areas are increasingly shaped by the big three, in which developments of decisive and immediate consequence now occur.

Families and churches and schools adapt to modern life; governments and armies and corporations shape it; and, as they do so, they turn these lesser institutions into means for their ends. Religious institutions provide chaplains to the armed forces where they are used as a means of increasing the effectiveness of its morale to kill. Schools select and train men for their jobs in corporations and their specialized tasks in the armed forces. The extended family has, of course, long been broken up by the industrial revolution, and now the son and the father are removed from the family, by compulsion if need be, whenever the army of the state sends out the call. And the symbols of all these lesser institutions are used to legitimate the power and the decisions of the big three.

The life-fate of the modern individual depends not only upon the family into which he was born or which he enters by marriage, but increasingly upon the corporation in which he spends the most alert hours of his best years; not only upon the school where he is educated as a child and adolescent, but also upon the state which touches him throughout his life; not only upon the church in which on occasion he hears the word of God, but also upon the army in which he is disciplined.

If the centralized state could not rely upon the inculcation of nationalist loyalties in public and private schools, its leaders would promptly seek to modify the decentralized educational system. If the bankruptcy rate among the top 500 corporations were as high as the general divorce rate among the 37 million married couples, there would be economic catastrophe on an international scale. If members of armies gave to them no more of their lives than do believers to the churches to which they belong, there would be a military crisis.

Within each of the big three, the typical institutional unit has become enlarged, has become administrative, and, in the power of its decisions, has become centralized. Behind these developments there is a fabulous technology, for as institutions, they have incorporated this technology and guide it, even as it shapes and paces their developments.

The economy—once a great scatter of small productive units in autonomous balance—has become dominated by two or three hundred giant corporations, administratively and politically interrelated, which together hold the keys to economic decisions.

The political order, once a decentralized set of several dozen states with a weak spinal cord, has become a centralized, executive establishment which has taken up into itself many powers previously scattered, and now enters into each and every cranny of the social structure.

The military order, once a slim establishment in a context of distrust fed by state militia, has become the largest and most expensive feature of government, and, although well-versed in smiling public relations, now has all the grim and clumsy efficiency of a sprawling bureaucratic domain.

In each of these institutional areas, the means of power at the disposal of decision makers have increased enormously; their central executive powers have been enhanced; within each of them modern administrative routines have been elaborated and tightened up.

As each of these domains becomes enlarged and centralized, the consequences of its activities become greater, and its traffic with the others increases. The decisions of a handful of corporations bear upon military and political as well as upon economic developments around the world. The decisions of the military establishment rest upon and grievously affect political life as well as the very level of economic activity. The decisions made within the political domain determine economic activities and military programs. There is no longer, on the one hand, an economy, and, on the other hand, a political order containing a military establishment unimportant to politics and to money-making. There is a political economy linked, in a thousand ways, with military institutions and decisions. On each side of the world-split running through central Europe and around the Asiatic rimlands, there is an ever-increasing inter-locking of economic, military, and political structures. If there is government intervention in the corporate economy, so is there corporate intervention in the governmental process. In the structural sense, this triangle of power is the source of the interlocking directorate that is most important for the historical structure of the present.

The fact of the interlocking is clearly revealed at each of the points of crisis of modern capitalist society—slump, war, and boom. In each, men of decision are led to an awareness of the interdependence of the major insitutional orders. In the nineteenth century, when the scale of all institutions was smaller, their liberal integration was achieved in the automatic economy, by an autonomous play of market forces, and in the automatic political domain, by the bargain and the vote. It was then assumed that out of the imbalance and friction that followed the limited decisions then possible a new equilibrium would in due course emerge. That can no longer be assumed, and it is not assumed by the men at the top of each of the three dominant hierarchies.

For given the scope of their consequences, decisions—and indecisions—Sany one of these ramify into the others, and hence top decisions tend either to become coordinated or to lead to a commanding indecision. It has not always been like this. When numerous small entrepreneurs made up the economy, for example, many of them could fail and the consequences still remain local; political and military authorities did not intervene. But now, given political expectations and military commitments, can they afford to allow

key units of the private corporate economy break down in [a] slump? Increasingly, they do intervene in economic affairs, and as they do so, the controlling decisions in each order are inspected by agents of the other two, and economic, military, and political structures are interlocked.

At the pinnacle of each of the three enlarged and centralized domains, there have arisen those higher circles which make up the economic, the political, and the military elites. At the top of the economy, among the corporate rich, there are the chief executives; at the top of the political order, the members of the political directorate; at the top of the military establishment, the elite of soldier-statesmen clustered in and around the Joint Chiefs of Staff and the upper echelon. As each of these domains has coincided with the others, as decisions tend to become total in their consequence, the leading men in each of the three domains of power—the warlords, the corporation chieftains, the political directorate—tend to come together, to form the power elite of America.

The higher circles in and around these command posts are often thought of in terms of what their members possess: They have a greater share than other people of the things and experiences that are most highly valued. From this point of view, the elite are simply those who have the most of what there is to have, which is generally held to include money, power, and prestige—as well as all the ways of life to which these lead. But the elite are not simply those who have the most, for they could not "have the most" were it not for their positions in the great institutions. For such institutions are the necessary bases of power, of wealth, and of prestige, and at the same time, the chief means of exercising power, of acquiring and retaining wealth, and of cashing in the higher claims for prestige.

By the powerful we mean, of course, those who are able to realize their will, even if others resist it. No one, accordingly, can be truly powerful unless he has access to the command of major institutions, for it is over these institutional means of power that the truly powerful are, in the first instance, powerful. Higher politicians and key officials of government command such institutional power; so do admirals and generals, and so do the major owners and executives of the larger corporations. Not all power, it is true, is anchored in and exercised by means of such institutions, but only within and through them can power be more or less continuous and important.

Wealth also is acquired and held in and through institutions. The pyramid of wealth cannot be understood merely in terms of the very rich; for the great inheriting families, as we shall see, are now supplemented by the coorporate institutions of modern society: Every one of the very rich families has been and is closely connected—always legally and frequently managerially as well—with one of the multimillion-dollar corporations.

The modern corporation is the prime source of wealth, but, in latter-day capitalism, the political apparatus also opens and closes many avenues to wealth. The amount as well as the source of income, the power over consumer's goods as well as over productive capital, are determined by position within the political economy. If our interest in the very rich goes beyond their lavish or their miserly consumption, we must examine their relations to modern forms of corporate property as well as to the state; for such relations now determine the chances of men to secure big property and to receive high income.

Great prestige increasingly follows the major institutional units of the social structure. It is obvious that prestige depends, often quite decisively, upon access to the publicity machines that are now a central and normal feature of all the big institutions of modern America. Moreover, one feature of these hierarchies of corporation, state, and military establishment is that their top positions are increasingly interchangeable. One result of this is the accumulative nature of prestige. Claims for prestige, for example, may be initially based on military roles, then expressed in and augmented by an educational institution run by corporate executives, and cashed in, finally, in the political order, where, for General Eisenhower and those he represents, power and prestige finally meet at the very peak. Like wealth and power, prestige tends to be cumulative: The more of it you have, the more you can get. These values also tend to be translatable into one another: The wealthy find it easier than the poor to gain power; those with status find it easier than those without it to control opportunities for wealth.

If we took the one-hundred most powerful men in America, the one-hundred wealthiest, and the one-hundred most celebrated away from the institutional positions they now occupy, away from their resources of men and women and money, away from the media of mass communication that are now focused upon them—then they would be powerless and poor and uncelebrated. For power is not of a man. Wealth does not center in the person of the wealthy. Celebrity is not inherent in any personality. To be celebrated, to be wealthy, to have power requires access to major institutions, for the institutional positions men occupy determine in large part their chances to have and to hold these valued experiences.

The people of the higher circles may also be conceived as members of a top social stratum, as a set of groups whose members know one another, see one another socially and at business, and so, in making decisions, take one another into account. The elite, according to this conception, feel themselves to be, and are felt by others to be, the inner circle of "the upper social classes." They form a more or less compact social and psychological entity; they have become self-conscious members of a social class. People are either accepted into this class or they are not, and there is a qualitative split, rather than merely a numerical scale, separating them from those who are not elite. They are more or less aware of themselves as a social class and they behave toward one another differently from the way they do toward members of other classes. They accept one another, understand one another, marry one another, tend to work and to think if not together at least alike.

Now, we do not want by our definition to prejudge whether the elite of the command posts are conscious members of such a socially recognized class, or whether considerable proportions of the elite derive from such a clear and distinct class. These are matters to be investigated. Yet in order to be able to recognize what we intend to investigate, we must note something that all biographies and memoirs of the wealthy and the powerful and the eminent make clear: No matter what else they may be, the people of these higher circles are involved in a set of overlapping "crowds" and intricately connected "cliques." There is a kind of mutual attraction among those who "sit on the same terrace"—although this often becomes clear to them, as well as to others, only at the point at which they feel the need to draw the line; only when, in their common defense, they come to understand what they have in common, and so close their ranks against outsiders.

The idea of such ruling stratum implies that most of its members have similar social origins, that throughout their lives they maintain a network of internal connections, and that to some degree there is an interchangeability of position between the various hierarchies of money and power and celebrity. We must, of course, note at once that if such an elite stratum does exist, its social visibility and its form, for very solid historical reasons, are quite different from those of the noble cousin-hoods that once ruled various European nations.

That American society has never passed through a feudal epoch is of decisive importance to the nature of the American elite, as well as to American society as a historic whole. For it means that no nobility or aristocracy, established before the capitalist era, has stood in tense opposition to the higher bourgeoisie. It means that this bourgeoisie has monopolized not only wealth but prestige and power as well. It means that no set of noble families has commanded the top positions and monopolized the values that are generally held in high esteem; and certainly that no set has done so explicitly by inherited right. It means that no high church dignitaries or court nobilities, no entrenched landlords with honorific accouterments, no monopolists of high army posts have opposed the enriched bourgeoisie and in the name of birth and prerogative successfully resisted its self-making.

But this does *not* mean that there are no upper strata in the United States. That they emerged from a "middle class" that had no recognized aristocratic superiors does not mean they remained middle class when enormous increases in wealth made their own superiority possible. Their origins and their newness may have made the upper strata less visible in America than elsewhere. But in America today there are in fact tiers and ranges of wealth and power of which people in the middle and lower ranks know very little and may not even dream. There are families who, in their well-being, are quite insulated from the economic jolts and lurches felt by the merely prosperous and those farther down the scale. There are also men of power who in quite small groups make decisions of enormous consequence for the underlying population. . . .

DISCUSSION QUESTIONS

1. Mills sees the power of certain institutions as having increased over time. What evidence do you see of the power of these institutions in your life?

2. The power elite, though composed of individuals, wields its influence because of its institutional control. How do these institutions interlock in the current period, and what consequences does this have for ordinary men and women?

3. What role does Mills see celebrities having with regard to the power elite? What evidence do you see of this in contemporary times?

INTERNET RESOURCES

Suggested Web URLs
for Further Study

*http://www.socialstudieshelp.com/APGOV_Power_
Elite.htm*
This Web site features an essay on The Power
Elite and provides specific examples that relate to
Mills' theories.

*http://www.angelfire.com/or/sociologyshop
/CWM.html*
A Web page with links to other articles by C.
Wright Mills and other essays commenting on
Mills and his works.

InfoTrac College Edition

You can find further relevant readings on the
World Wide Web at
http://sociology.wadsworth.com

Virtual Society

For further information on this subject including
links to relevant Web sites, go to the Wadsworth
Sociology homepage at
http://sociology.wadsworth.com

14

Savage Inequalities

JONATHAN KOZOL

"East of anywhere," writes a reporter for the *St. Louis Post-Dispatch,* "often evokes the other side of the tracks. But, for a first-time visitor suddenly deposited on its eerily empty streets, East St. Louis might suggest another world." The city, which is 98 percent black, has no obstetric services, no regular trash collection, and few jobs. Nearly a third of its families live on less than $7,500 a year: 75 percent of its population lives on welfare of some form. The U.S. Department of Housing and Urban Development describes it as "the most distressed small city in America."

Only three of the 13 buildings on Missouri Avenue, one of the city's major thoroughfares, are occupied. A 13-story office building, tallest in the city, has been boarded up. Outside, on the sidewalk, a pile of garbage fills a ten-foot crater.

The city, which by night and day is clouded by the fumes that pour from vents and smokestacks at the Pfizer and Monsanto chemical plants, has one of the highest rates of child asthma in America.

It is, according to a teacher at Southern Illinois University, "a repository for a nonwhite population that is now regarded as expendable." The *Post-Dispatch* describes it as "America's Soweto."

Fiscal shortages have forced the layoff of 1,170 of the city's 1,400 employees in the past 12 years. The city, which is often unable to buy heating fuel or toilet paper for the city hall, recently announced that it might have to cashier all but 10 percent of the remaining work force of 230. In 1989 the mayor announced that he might need to sell the city hall and all six fire stations to raise needed cash. Last year the plan had to be scrapped after the city lost its city hall in a court judgment to a creditor. East St. Louis is mortgaged into the next century but has the highest property-tax rate in the state. . . .

The dangers of exposure to raw sewage, which backs up repeatedly into the homes of residents in East St. Louis, were first noticed, in the spring of 1989, at a public housing project, Villa Griffin. Raw sewage, says the *Post-Dispatch,* overflowed into a playground just behind the housing project, which is home to 187 children, "forming an oozing lake of . . . tainted water." . . . . A St. Louis health official voices her dismay that children live with waste in their backyards. "The development of working sewage systems made cities livable a hundred years ago," she notes. "Sewage systems separate us from the Third World." . . .

The sewage, which is flowing from collapsed pipes and dysfunctional pumping stations, has also flooded basements all over the city. The city's vacuum truck, which uses water and suction to unclog the city's sewers, cannot be used because it needs $5,000 in repairs. Even when it works, it sometimes can't be used because there isn't money to hire drivers. A single engineer now does the work that 14 others did before they were laid off. By April the pool of overflow behind the Villa Griffin project has expanded into a lagoon of sewage. Two million gallons of raw sewage lie outside the children's homes. . . .

. . . Sister Julia Huiskamp meets me on King Boulevard and drives me to the Griffin homes.

As we ride past blocks and blocks of skeletal structures, some of which are still inhabited, she slows the car repeatedly at railroad crossings. A seemingly endless railroad train rolls past us to the right. On the left: a blackened lot where garbage has been burning. Next to the burning garbage is a row of 12 white cabins, charred by fire. Next: a lot that holds a heap of auto tires and a mountain of tin cans. More burnt houses. More trash fires. The train moves almost imperceptibly across the flatness of the land.

Fifty years old, and wearing a blue suit, white blouse, and blue headcover, Sister Julia points to the nicest house in sight. The sign on the front reads MOTEL. "It's a whorehouse," Sister Julia says.

When she slows the car beside a group of teen-age boys, one of them steps out toward the car, then backs away as she is recognized.

The 99 units of the Villa Griffin homes— two-story structures, brick on the first floor, yellow wood above—form one border of a recessed park and playground that were filled with fecal matter last year when the sewage mains exploded. The sewage is gone now and the grass is very green and looks inviting. When nine-year-old Serena and her seven-year-old brother take me for a walk, however, I discover that our shoes sink into what is still a sewage marsh. An inch-deep residue of fouled water still remains.

Serena's brother is a handsome, joyous little boy, but troublingly thin. Three other children join us as we walk along the marsh: Smokey, who is none years old but cannot yet tell time; Mickey, who is seven; and a tiny child with a ponytail and big brown eyes who talks a constant stream of words that I can't always understand.

"Hush, Little Sister," says Serena. I ask for her name, but "Little Sister" is the only name the children seem to know.

"There go my cousins," Smokey says, pointing to two teen-age girls above us on the hill.

The day is warm, although we're only in the second week of March; several dogs and cats are playing by the edges of the marsh. "It's a lot of squirrels here," says Smokey. "There go one!"

"This here squirrel is a friend of mine," says Little Sister.

None of the children can tell me the approximate time that school begins. One says five o'clock. One says six. Another says that school begins at noon.

When I ask what song they sing after the flag pledge, one says, "Jingle Bells."

Smokey cannot decide if he is in the second or third grade.

Seven-year-old Mickey sucks his thumb during the walk.

The children regale me with a chilling story as we stand beside the marsh. Smokey says his sister was raped and murdered and then dumped behind his school. Other children add more details: Smokey's sister was 11 years old. She was beaten with a brick until she died. The murder was committed by a man who knew her mother.

The narrative begins when, without warning, Smokey says, "My sister has got killed."

"She was my best friend," Serena says.

"They had beat her in the head and raped her," Smokey says.

"She was hollering out loud," says Little Sister.

I ask them when it happened. Smokey says, "Last year." Serena then corrects him and she says, "Last week."

"It scared me because I had to cry," says Little Sister.

"The police arrested one man but they didn't catch the other," Smokey says.

Serena says, "He was some kin to her."

But Smokey objects, "He weren't no kin to me. He was my momma's friend." "Her face was busted," Little Sister says.

Serena describes this sequence of events: "They told her go behind the school. They'll give her a quarter if she do. Then they knock her down and told her not to tell what they had did."

I ask, "Why did they kill her?"

"They was scared that she would tell," Serena says.

"One is in jail," says Smokey. "They cain't find the other."

"Instead of raping little bitty children, they should find themselves a wife," says Little Sister.

"I hope," Serena says, "her spirit will come back and get that man."

"And *kill* that man," says Little Sister.

"Give her another chance to live," Serena says.

"My teacher came to the funeral," says Smokey.

"When a little child dies, my momma say a star go straight to Heaven," says Serena.

"My grandma was murdered," Mickey says out of the blue. "Somebody shot two bullets in her head."

I ask him, "Is she really dead?"

"She dead all right," says Mickey. "She was layin' there, just dead."

"I love my friends," Serena says. "I don't care if they no kin to me. I care for them. I hope his mother have another baby. Name her for my friend that's dead."

"I have a cat with three legs," Smokey says.

"Snakes hate rabbits," Mickey says, again for no apparent reason.

"Cats hate fishes," Little Sister says.

"It's a lot of hate," says Smokey.

Later, at the mission, Sister Julia tells me this: "The Jefferson School, which they attend, is a decrepit hulk. Next to it is a modern school, erected two years ago, which was to have replaced the one that they attend. But the construction was not done correctly. The roof is too heavy for the walls, and the entire structure has begun to sink. It can't be occupied. Smokey's sister was raped and murdered and dumped between the old school and the new one.". . .

The problems of the streets in urban areas, as teachers often note, frequently spill over into public schools. In the public schools of East St. Louis this is literally the case.

"Martin Luther King Junior High School," notes the *Post-Dispatch* in a story published in the early spring of 1989, "was evacuated Friday afternoon after sewage flowed into the kitchen. . . . The kitchen was closed and students were sent home." On Monday, the paper continues, "East St. Louis Senior High School was awash in sewage for the second time this year." The school had to be shut because of "fumes and backed-up toilets." Sewage flowed into the basement, through the floor, then up into the kitchen and the students' bathrooms. The backup, we read, "occurred in the food preparation areas."

School is resumed the following morning at the high school, but a few days later the overflow recurs. This time the entire system is affected, since the meals distributed to every student in the city are prepared in the two schools that have been flooded. School is called off for all 16,500 students in the district. The sewage backup, caused by the failure of two pumping stations, forces officials at the high school to shut down the furnaces.

At Martin Luther King, the parking lot and gym are also flooded. "It's a disaster," says a legislator. "The streets are under water; gaseous fumes are being emitted from the pipes under the schools," she says, "making people ill."

In the same week, the schools announce the layoff of 280 teachers, 166 cooks and cafeteria workers, 25 teacher aides, 16 custodians and 18 painters, electricians, engineers and plumbers. The president of the teachers' union says the cuts, which will bring the size of kindergarten and primary classes up to 30 students, and the size of fourth to twelfth grade classes up to 35, will have "an unimaginable impact" on the students. "If you have a high school teacher with five classes each day and between 150 and 175 students . . ., it's going to have a devastating effect." The school system, it is also noted, has been using more than 70 "permanent substitute teachers," who are paid only $10,000 yearly, as a way of saving money. . . .

East St. Louis, says the chairman of the state board, "is simply the worst possible place I can imagine to have a child brought up. . . . The community is in desperate circumstances." Sports and music, he observes, are, for many children here, "the only avenues of success." Sadly enough, no matter how it ratifies the stereotype, this is the truth; and there is a poignant aspect to the fact that, even with class size soaring and one quarter of the system's teachers being given their dismissal, the state board of education demonstrates its genuine but skewed compassion by attempting to leave sports and music untouched by the overall austerity.

Even sports facilities, however, are degrading by comparison with those found and expected at most high schools in America. The football field at East St. Louis High is missing almost everything—including goalposts. There are a couple of metal pipes—no crossbar, just the pipes. Bob Shannon, the football coach, who has to use his personal funds to purchase footballs and has had to cut and rake the football field himself, has dreams of having goalposts someday. He'd also like to let his students have new uniforms. The ones they wear are nine years old and held together somehow by a patchwork of repairs. Keeping them clean is a problem, too. The school cannot afford a washing machine. The uniforms are carted to a corner laundromat with fifteen dollars' worth of quarters. . . .

In the wing of the school that holds vocational classes, a damp, unpleasant odor fills the halls. The school has a machine shop, which cannot be used for lack of staff, and a woodworking shop. The only shop that's occupied this morning is the auto-body class. A man with long blond hair and wearing a white sweat suit swings a paddle to get children in their chairs. "What we need the most is new equipment," he reports. "I have equipment for alignment, for example, but we don't have money to install it. We also need a better form of egress. We bring the cars in through two other classes." Computerized equipment used in most repair shops, he reports, is far beyond the high school's budget. It looks like a very old gas station in an isolated rural town. . . .

The science labs at East St. Louis High are 30 to 50 years outdated. John McMillan, a soft-spoken man, teaches physics at the school. He shows me his lab. The six lab stations in the room have empty holes where pipes were once attached. "It would be great if we had water," says McMillan. . . .

Leaving the chemistry labs, I pass a double-sized classroom in which roughly 60 kids are sitting fairly still but doing nothing. "This is supervised study hall," a teacher tells me in the corridor. But when we step inside, he finds there is no teacher. "The teacher must be out today," he says.

Irl Solomon's history classes, which I visit next, have been described by journalists who cover East St. Louis as the highlight of the school. Solomon, a man of 54 whose reddish hair is turning white, has taught in urban schools for almost 30 years. A graduate of Brandeis University, he entered law school but was drawn away by a concern with civil rights. "After one semester, I decided that the law was not for me. I said, 'Go and find the toughest place there is to teach. See if you like it.' I'm still here. . . .

"I have four girls right now in my senior home room who are pregnant or have just had babies. When I ask them why this happens, I am told, 'Well, there's no reason not to have a baby. There's not much for me in public school.' The truth is, that's a pretty honest answer. A diploma from ghetto high school doesn't count for much in the United States today. So, if this is really the last education that a person's going to get, she's probably perceptive in that statement. Ah, there's so much bitterness—unfairness—there, you know. Most of these pregnant girls are not the ones who have much self-esteem. . . .

"Very little education in the school would be considered academic in the suburbs. Maybe 10 to 15 percent of students are in truly academic programs. Of the 55 percent who graduate, 20 percent may go to four-year colleges: something like 10 percent of any entering class. Another 10 to 20 percent may get some other kind of higher education. An equal number join the military. . . .

"I don't go to physics class, because my lab has no equipment," says one student. "The typewriters in my typing class don't work. The women's toilets . . ." She makes a sour face. "I'll be honest," she says. "I just don't use the toilets. If I do, I come back into class and I feel dirty."

"I wanted to study Latin," says another student. "But we don't have Latin in this school."

"We lost our only Latin teacher," Solomon says.

A girl in a white jersey with the message DO THE RIGHT THING on the front raises her hand. "You visit other schools," she says. "Do you think the children in this school are getting what we'd get in a nice section of St. Louis?"

I note that we are in a different state and city.

"Are we citizens of East St. Louis or America?" she asks. . . .

In a seventh grade social studies class, the only book that bears some relevance to black concerns—its title is *The American Negro*—bears a publication date of 1967. The teacher invites me to ask the class some questions. Uncertain where to start, I ask the students what they've learned about the civil rights campaigns of recent decades.

A 14-year-old girl with short black curly hair says this: "Every year in February we are told to read the same old speech of Martin Luther King. We read it every year. 'I have a dream. . . .' It does begin to seem—what is the word?" She hesitates and then she finds the word: "perfunctory."

I ask her what she means.

"We have a school in East St. Louis named for Dr. King," she says. "The school is full of sewer water and the doors are locked with chains. Every student in that school is black. It's like a terrible joke on history."

It startles me to hear her words, but I am startled even more to think how seldom any press reporter has observed the irony of naming segregated schools for Martin Luther King. Children reach the heart of these hypocrisies much quicker than the grown-ups and the experts do. . . .

★ ★ ★

The train ride from Grand Central Station to suburban Rye, New York, takes 35 to 40 minutes. The high school is a short ride from the station. Built of handsome gray stone and set in a landscaped campus, it resembles a New England prep school. On a day in early June of 1990, I enter the school and am directed by a student to the office.

The principal, a relaxed, unhurried man who, unlike many urban principals, seems gratified to have me visit in his school, takes me in to see the auditorium, which, he says, was recently restored with private charitable funds ($400,000) raised by parents. The crenellated ceiling, which is white and spotless, and the polished dark-wood paneling contrast with the collapsing structure of the auditorium at [another school I visited]. The principal strikes his fist against the balcony: "They made this place extremely solid." Through a window, one can see the spreading branches of a beech tree in the central courtyard of the school.

In a student lounge, a dozen seniors are relaxing on a carpeted floor that is constructed with a number of tiers so that, as the principal explains, "they can stretch out and be comfortable while reading."

The library is wood-paneled, like the auditorium. Students, all of whom are white, are seated at private carrels, of which there are approximately 40. Some are doing homework; others are looking through the *New York Times*. Every student that I see during my visit to the school is white or Asian, though I later learn there are a number of Hispanic students and that 1 or 2 percent of students in the school are black.

According to the principal, the school has 96 computers for 546 children. The typical student, he says, studies a foreign language for four or five years, beginning in the junior high school, and a second foreign language (Latin is available) for two years. Of 140 seniors, 92 are now enrolled in AP [advanced placement] class. Maximum teacher salary will soon reach $70,000. Per pupil funding is above $12,000 at the time I visit.

The students I meet include eleventh and twelfth graders. The teacher tells me that the class is reading Robert Coles, Studs Terkel, Alice Walker. He tells me I will find them more than willing to engage me in debate, and this turns out to be correct. Primed for my visit, it appears, they arrow in directly on the dual questions of equality and race.

Three general positions soon emerge and seem to be accepted widely. The first is that the fiscal inequalities "do matter very much" in shaping what a school can offer ("That is obvious," one student says) and that any loss of funds in Rye, as a potential consequence of future equalizing, would be damaging to many things the town regards as quite essential.

The second position is that racial integration—for example, by the busing of black children from the city or a nonwhite suburb to this school—would meet with strong resistance, and the reason would not simply be the fear that certain standards might decline. The reason, several students say straightforwardly, is "racial" or, as others say it, "out-and-out racism" on the part of adults.

The third position voiced by many students, but not all, is that equity is basically a goal to be desired and should be pursued for moral reasons, but "will probably make no major difference" since poor children "still would lack the motivation" and "would probably fail in any case because of other problems."

At this point, I ask if they can truly say "it wouldn't make a difference" since it's never been attempted. Several students then seem to rethink their views and says that "it might work, but it would have to start with preschool and the elementary grades" and "it might be 20 years before we'd see a difference."

At this stage in the discussion, several students speak with some real feeling of the present inequalities, which, they say, are "obviously unfair," and one student goes a little further and proposes that "we need to change a lot more than the schools." Another says she'd favor racial integration "by whatever means—including busing— even if the parents disapprove." But a contradictory opinion also is expressed with a good deal of fervor and is stated by one student in a rather biting voice: "I don't see why we should do it. How could it be of benefit to us?"

Throughout the discussion, whatever the views the children voice, there is a degree of unreality about the whole exchange. The children are lucid and their language is well chosen and their arguments well made, but there is a sense that they are dealing with an issue that does not feel very vivid, and that nothing that we say about it to each other really matters since it's "just a theoretical discussion." To a certain degree, the skillfulness and cleverness that they display seem to derive precisely from this sense of unreality. Questions of unfairness feel more like a geometric problem than a matter of humanity or conscience. A few of the students do break through the note of unreality, but, when they do, they cease to be so agile in their use of words and speak more awkwardly. Ethical challenges seem to threaten their effectiveness. There is the sense that they were skating over ice and that the issues we addressed were safely frozen underneath. When they stop to look beneath the ice they start to stumble. The verbal competence they have acquired here may have been gained by building walls around some regions of the heart.

"I don't think that busing students from their ghetto to a different school would do much good," one student says. "You can take them out of the environment, but you can't take the environment out of *them*. If someone grows up in the South Bronx, he's not going to be prone to learn." His name is Max and he has short black hair and speaks with confidence. "Busing didn't work when it was tried," he says. I ask him how he knows this and he says he saw a television movie about Boston.

"I agree that it's unfair the way it is," another student says. "We have AP courses and they don't. Our classes are much smaller." But, she says, "putting them in schools like ours is not the answer. Why not put some AP classes into *their* school? Fix the roof and paint the halls so it will not be so depressing."

The students know the term "separate but equal," but seem unaware of its historical associations. "Keep them where they are but make it equal," says a girl in the front row.

A student named Jennifer, whose manner of speech is somewhat less refined and polished than that of the others, tells me that her parents came here from New York. "My family is originally from the Bronx. Schools are hell there. That's one reason that we moved. I don't think it's our responsibility to pay our taxes to provide for them. I mean, my parents used to live there and they wanted to get out. There's no point in coming to a place like this, where schools are good, and then your taxes go back to the place where you began."

I bait her a bit: "Do you mean that, now that you are not in hell, you have no feeling for the people that you left behind?"

"It has to be the people in the area who want an education. If your parents just don't care, it won't do any good to spend a lot of money. Someone else can't want a good life for you. You have got to want it for yourself." Then she adds, however, "I agree that everyone should have a chance at taking the same courses. . . ."

I ask her if she'd think it fair to pay more taxes so that this was possible.

"I don't see how that benefits me," she says.

DISCUSSION QUESTIONS

1. Inequality permeates educational institutions, as Kozol's essay shows. What specific inequalities in education does he note? Do you see these reflected in the school in your community?

2. How does Kozol's description of urban schools show a connection between race and class inequality?

3. One way that sociologists think about education is as a route to social mobility. Others see education as only reproducing the inequalities that exist in society. As you observe education today, what evidence do you see for each of these perspectives on the role of education in society?

INTERNET RESOURCES

Suggested Web URLs
for Further Study

http://www.nwrel.org/charter/policy.html
This Web site on charter schools details attempts to reform schools and reach toward equality through charter school initiatives.

http://www.nextcity.com/main/town/4editor.htm #baster
An editorial and responses out of Canada about poverty. A very interesting article bound to spark debate.

http://www.ed.gov/NCES/index.html
Official site for the National Center for Education Statistics (NCES). This site features information on literacy, education costs, student assessment, and more in the United States and around the world.

InfoTrac College Edition

You can find further relevant readings on the World Wide Web at
http://sociology.wadsworth.com

Virtual Society

For further information on this subject including links to relevant Web sites, go to the Wadsworth Sociology homepage at
http://sociology.wadsworth.com

15

Urbanism as a Way of Life

LOUIS WIRTH

A SOCIOLOGICAL DEFINITION OF THE CITY

Despite the preponderant significance of the city in our civilization, our knowledge of the nature of urbanism and the process of urbanization is meager, notwithstanding many attempts to isolate the distinguishing characteristics of urban life. Geographers, historians, economists, and political scientists have incorporated the points of view of their respective disciplines into diverse definitions of the city. While in no sense intended to supersede these, the formulation of a sociological approach to the city may incidentally serve to call attention to the interrelations between them by emphasizing the peculiar characteristics of the city as a particular form of human association. A sociologically significant definition of the city seeks to select those elements of urbanism which mark it as a distinctive mode of human group life. . . .

For sociological purposes a city may be defined as a relatively large, dense, and permanent settlement of socially heterogeneous individuals. On the basis of the postulates which this minimal definition suggests, a theory of urbanism may be formulated in the light of existing knowledge concerning social groups.

A THEORY OF URBANISM

Given a limited number of identifying characteristics of the city, I can better assay the consequences or further characteristics of them in the light of general sociological theory and empirical research. I hope in this manner to arrive at the essential propositions comprising a theory of urbanism. Some of these propositions can be supported by a considerable body of already available research materials; others may be accepted as hypotheses for which a certain amount of presumptive evidence exists, but for which more ample and exact verification would be required. At least such a procedure will, it is hoped, show what in the way of systematic knowledge of the city we now have and what are the crucial and fruitful hypotheses for future research.

The central problem of the sociologist of the city is to discover the forms of social action and organization that typically emerge in relatively permanent, compact settlements of large numbers

Credit: "Urbanism as a Way of Life," by Louis Wirth from *American Journal of Sociology,* Vol. 44, no.1, July 1938, pp. 1-24. Reprinted by permission of the University of Chicago Press.

of heterogeneous individuals. We must also infer that urbanism will assume its most characteristic and extreme form in the measure in which the conditions with which it is congruent are present. Thus the larger, the more densely populated, and the more heterogeneous a community, the more accentuated the characteristics associated with urbanism will be. . . .

Some justification may be in order for the choice of the principal terms comprising our definition of the city, a definition which ought to be as inclusive and at the same time as denotative as possible without unnecessary assumptions. To say that large numbers are necessary to constitute a city means, of course, large numbers in relation to a restricted area or high density of settlement. There are, nevertheless, good reasons for treating large numbers and density as separate factors, because each may be connected with significantly different social consequences. Similarly the need for adding heterogeneity to numbers of population as a necessary and distinct criterion of urbanism might be questioned, since we should expect the range of differences to increase with numbers. In defense, it may be said that the city shows a kind and degree of heterogeneity of population which cannot be wholly accounted for by the law of large numbers or adequately represented by means of a normal distribution curve. Because the population of the city does not reproduce itself, it must recruit its migrants from other cities, the countryside, and—in the United States until recently—from other countries. The city has thus historically been the melting-pot of races, peoples, and cultures, and a most favorable breeding-ground of new biological and cultural hybrids. It has not only tolerated but rewarded individual differences. It has brought together people from the ends of the earth because they are different and thus useful to one another, rather than because they are homogeneous and like-minded.

A number of sociological propositions concerning the relationship between (a) numbers of population, (b) density of settlement, (c) heterogeneity of inhabitants and group life can be formulated on the basis of observation and research.

Size of the Population Aggregate

Ever since Aristotle's *Politics,* it has been recognized that increasing the number of inhabitants in a settlement beyond a certain limit will affect the relationships between them and the character of the city. Large numbers involve, as has been pointed out, a greater range of individual variation. Furthermore, the greater the number of individuals participating in a process of interaction, the greater is the *potential* differentiation between them. The personal traits, the occupations, the cultural life, and the ideas of the members of an urban community may, therefore, be expected to range between more widely separated poles than those of rural inhabitants.

That such variations should give rise to the spatial segregation of individuals according to color, ethnic heritage, economic and social status, tastes and preferences, may readily be inferred. The bonds of kinship, of neighborliness, and the sentiments arising out of living together for generations under a common folk tradition are likely to be absent or, at best, relatively weak in an aggregate the members of which have such diverse origins and backgrounds. Under such circumstances competition and formal control mechanisms furnish the substitutes for the bonds of solidarity that are relied upon to hold a folk society together.

Increase in the number of inhabitants of a community beyond a few hundred is bound to limit the possibility of each member of the community knowing all the others personally. Max Weber, in recognizing the social significance of this fact, explained that from a sociological point of view large numbers of inhabitants and density of settlement mean a lack of that mutual acquaintanceship which ordinarily inheres between the inhabitants in a neighborhood.[1] The increase in numbers thus involves a changed character of the social relationships. As Georg Simmel points out: "[If] the unceasing external contact of numbers of persons in the city should be met by the same number of inner reactions as in the small town, in which one knows almost every person he meets and to each of whom he has a positive relationship, one would be completely atomized internally and would fall into an unthinkable mental condition."[2] The multiplication of persons in a state of interaction under conditions which make their contact as full personalities impossible produces that segmentalization of human relationships which has sometimes been seized upon by students of the mental life of the cities as an explanation for the "schizoid" character of urban personality. This is not to say that the urban inhabitants have fewer acquaintances than rural inhabitants, for the reverse may actually be true; it means rather that in relation to the number of people whom they see and with whom they rub elbows in the course of daily life, they know a smaller proportion, and of these they have less intensive knowledge.

Characteristically, urbanites meet one another in highly segmental roles. They are, to be sure, dependent upon more people for the satisfactions of their life-needs than are rural people and thus are associated with a greater number of organized groups, but they are less dependent upon particular persons, and their dependence upon others is confined to a highly fractionalized aspect of the other's round of activity. This is essentially what is meant by saying that the city is characterized by secondary rather than primary contacts. The contacts of the city may indeed be face to face, but they are nevertheless impersonal, superficial, transitory, and segmental. The reserve, the indifference, and the blasé outlook which urbanites manifest in their relationships may thus be regarded as devices for immunizing themselves against the personal claims and expectations of others.

The superficiality, the anonymity, and the transitory character of urban social relations make intelligible, also, the sophistication and the rationality generally ascribed to city-dwellers. Our acquaintances tend to stand in a relationship of utility to us in the sense that the role which each one plays in our life is overwhelmingly regarded as a means for the achievement of our own ends. Whereas the individual gains, on the one hand, a certain degree of emancipation or freedom from the personal and emotional controls of intimate groups, he loses, on the other hand, the spontaneous self-expression, the morale, and the sense of participation that comes with living in an integrated society. This constitutes essentially the state of *anomie,* or the social void, to which Durkheim alludes in attempting to account for the various forms of social disorganization in technological society.

The segmental character and utilitarian accent of interpersonal relations in the city find their institutional expression in the proliferation of specialized tasks which we see in their most developed form in the professions. The operations of the pecuniary nexus lead to predatory relationships, which tend to obstruct the efficient functioning of the social order unless checked by professional codes and occupational etiquette. The premium put upon utility and efficiency suggests the adaptability of the corporate device for the organization of enterprises in which individuals can engage only in groups. The advantage that the corporation has over the individual entrepreneur and the partnership in the urban–industrial world derives not only from the possibility it affords of centralizing the resources of thousands of individuals or from the legal privilege of limited liability and perpetual succession, but from the fact that the corporation has no soul.

The specialization of individuals, particularly in their occupations, can proceed only, as Adam Smith pointed out, upon the basis of an enlarged market, which in turn accentuates the division of labor. This enlarged market is only in part supplied by the city's hinterland; in large measure it is found among the large numbers that the city itself contains. The dominance of the city over the surrounding hinterland becomes explicable in terms of the division of labor which urban life occasions and promotes. The extreme degree of interdependence and the unstable equilibrium of urban life are closely associated with the division of labor and the specialization of occupations. This interdependence and this instability are increased by the tendency of each city to specialize in those functions in which it has the greatest advantage.

In a community composed of a larger number of individuals that can know one another intimately and can be assembled in one spot, it becomes necessary to communicate through indirect media and to articulate individual interests by a process of delegation. Typically in the city, interests are made effective through representation. The individual counts for little, but the voice of the representative is heard with a deference roughly proportional to the numbers for whom he speaks.

While this characterization of urbanism, in so far as it derives from large numbers, does not by any means exhaust the sociological inferences that might be drawn from our knowledge of the relationship of the size of a group to the characteristic behavior of the members, for the sake of brevity the assertions made may serve to exemplify the sort of propositions that might be developed.

Density

As in the case of numbers, so in the case of concentration in limited space certain consequences of relevance in sociological analysis of the city emerge. Of these only a few can be indicated.

As Darwin pointed out for flora and fauna and as Durkheim noted in the case of human societies,[3] an increase in numbers when area is held constant (i.e., an increase in density) tends to produce differentiation and specialization, since only in this way can the area support increased numbers. Density thus reinforces the effect of numbers in diversifying men and their activities and in increasing the complexity of the social structure.

On the subjective side, as Simmel has suggested, the close physical contact of numerous individuals necessarily produces a shift in the media through which we orient ourselves to the urban milieu, especially to our fellow-men. Typically, our physical contacts are close but our social contacts are distant. The urban world puts a premium on visual recognition. We see the uniform which denotes the role of the functionaries, and are oblivious to the personal eccentricities hidden behind the uniform. We tend to acquire and develop a sensitivity to a world of artifacts, and become progressively farther removed from the world of nature.

We are exposed to glaring contrasts between splendor and squalor, between riches and poverty, intelligence and ignorance, order and chaos. The competition for space is great, so that each area generally tends to be put to the use which yields the greatest economic return. Place of work tends to become dissociated from place of residence, for the proximity of industrial and commercial establishments makes an area both economically and socially undesirable for residential purposes.

Density, land values, rentals, accessibility, healthfulness, prestige, aesthetic consideration, absence of nuisances such as noise, smoke, and dirt determine the desirability of various areas of the city as places of settlement for different sections of the population. Place and nature of work, income, racial and ethnic characteristics, social status, custom, habit, taste, preference, and prejudice are among the significant factors in accordance with which the urban population is selected and distributed into more or less distinct settlements. Diverse population elements inhabiting a compact settlement thus become segregated from one another in the degree in which their requirements and modes of life are incompatible and in the measure in which they are antagonistic. Similarly, persons of homogeneous status and needs unwittingly drift into, consciously select, or are forced by circumstances into the same area. The different parts of the city acquire specialized functions, and the city consequently comes to resemble a mosaic of social worlds in which the transition from one to the other is abrupt. The juxtaposition of divergent personalities and modes of life tends to produce a relativistic perspective and a sense of toleration of differences which may be regarded as prerequisites for rationality and which lead toward the secularization of life.[4]

The close living together and working together of individuals who have no sentimental and emotional ties foster a spirit of competition, aggrandizement, and mutual exploitation. Formal controls are instituted to counteract irresponsibility and potential disorder. Without rigid adherence to predictable routines a large compact society would scarcely be able to maintain itself. The clock and the traffic signal are symbolic of the basis of our social order in the urban world. Frequent close physical contact, coupled with great social distance, accentuates the reserve of unattached individuals toward one another and, unless compensated by other opportunities for response, gives rise to loneliness. The necessary frequent movement of great numbers of individuals in a congested habitat causes friction and irritation. Nervous tensions which derive from such personal frustrations are increased by the rapid tempo and the complicated technology under which life in dense areas must be lived.

Heterogeneity

The social interaction among such a variety of personality types in the urban milieu tends to break down the rigidity of caste lines and to complicate the class structure; it thus induces a more ramified and differentiated framework of social stratification than is found in more integrated societies. The heightened mobility of the individual, which brings him within the range of stimulation by a great number of diverse individuals and subjects him to fluctuating status in the differentiated social groups that compose the social structure of the city, brings him toward the acceptance of instability and insecurity in the world at large as a norm. This fact helps to account, too, for the sophistication and cosmopolitanism of the urbanite. No single group has the undivided allegiance of the individual. The groups with which he is affiliated do not lend themselves readily to a simple hierarchical arrangement. By virtue of his different interests arising out of different aspects of social life, the individual acquires membership in widely divergent groups, each of which functions only with reference to a single segment of his personality.

Nor do these groups easily permit of a concentric arrangement so that the narrower ones fall within the circumference of the more inclusive ones, as is more likely to be the case in the rural community or in primitive societies. Rather the groups with which the person typically is affiliated are tangential to each other or intersect in highly variable fashion.

Partly as a result of the physical footlooseness of the population and partly as a result of their social mobility, the turnover in group membership generally is rapid. Place of residence, place and character of employment, income, and interests fluctuate, and the task of holding organizations together and maintaining and promoting intimate and lasting acquaintanceship between the members is difficult. This applies strikingly to the local areas within the city into which persons become segregated more by virtue of differences in race, language, income, and social status than through choice or positive attraction to people like themselves. Overwhelmingly the city-dweller is not a home-owner, and since a transitory habitat does not generate binding traditions and sentiments, only rarely is he a true neighbor. There is little opportunity for the individual to obtain a conception of the city as a whole or to survey his place in the total scheme. Consequently he finds it difficult to determine what is to his own "best interests" and to decide between the issues and leaders presented to him by the agencies of mass suggestion. Individuals who are thus detached from the organized bodies which integrate society comprise the fluid masses that make collective behavior in the urban community to unpredictable and hence so problematical.

Although the city, through the recruitment of variant types to perform its diverse tasks and the accentuation of their uniqueness through competition and the premium upon eccentricity, novelty, efficient performance, and inventiveness, produces a highly differentiated population, it also exercises a leveling influence. Wherever large numbers of differently constituted individuals congregate, the process of depersonalization also enters. This leveling tendency inheres in part in the economic basis of the city. The development of large cities, at least in the modern age, was largely dependent upon the concentrative force of steam. The rise of the factory made possible mass production for an impersonal market. The fullest exploitation of the possibilities of the division of labor and mass production, however, is possible only with standardization of processes and products. A money economy goes hand in hand with such a system of production. Progressively as cities have developed upon a background of this system of production, the pecuniary nexus which implies the purchasability of services and things has displaced personal relations as the basis of association. Individuality under these circumstances must be replaced by categories. When large numbers have to make common use of facilities and institutions, those facilities and institutions must serve the needs of the average person rather than those of particular individuals. The services of the public utilities, of the recreational, educational, and cultural institutions, must be adjusted to mass requirements. Similarly, the cultural institutions, such as the schools, the movies, the radio, and the newspapers, by virtue of their mass clientele, must necessarily operate as leveling influences. The political process as it appears in urban life could not be understood unless one examined the mass appeals made through modern propaganda techniques. If the individual would participate at all in the social, political, and economic life of the city, he must subordinate some of his individuality to the demands of the larger community and in that measure immerse himself in mass movements. . . .

On the basis of the three variables, number, density of settlement, and degree of heterogeneity, of the urban population, it appears possible to explain the characteristics of urban life and to account for the differences between cities of various sizes and types. . . .

NOTES

1. *Wirtschaft und Gesellschaft* (Tübingen, 1925), part I, chap. 8, p. 514.

2. "Die Grossstädte und das Geistesleben," *Die Grosstadt,* ed. Theodor Petermann (Dresden, 1903). pp. 187–206.

3. E. Durkheim, *De la division du travail socia* (Paris, 1932), p. 248.

4. The extent to which the segregation of the population into distinct ecological and cultural areas and the resulting social attitude of tolerance, rationality, and secular mentality are functions of density as distinguished from heterogeneity is difficult to determine. Most likely we are dealing here with phenomena which are consequences of the simultaneous operation of both factors.

DISCUSSION QUESTIONS

1. What are the factors about city life that Wirth associates with specific forms of behavior? How does each factor specifically affect the behavior he calls urbanism?

2. Since Wirth's essay was written, much of U.S. society has become suburbanized. What factors do you think Wirth would not associate with suburbanism and how would these factors influence social behavior?

3. Do you think that different cities might have different forms of urbanism because of variations in the particular characteristics of a given city? What characteristics would matter and how might they influence the form of behavior that Wirth identifies as urbanism? (Hint: Think of things such as the economic base of the city, the racial-ethnic population, region, etc.).

INTERNET RESOURCES

Suggested Web URLs
for Further Study

http://www.emich.edu/public/geo/rtownintro.html
Rivertown Introduction: Urban Computer Simulation Project (Urban and Regional Planning at Eastern Michigan University)

http://www.census.gov/ftp/pub/population/www/
Population Division Home Page

InfoTrac College Edition

You can find further relevant readings on the World Wide Web at
http://sociology.wadsworth.com

Virtual Society

For further information on this subject including links to relevant Web sites, go to the Wadsworth Sociology homepage at
http://sociology.wadsworth.com

Glossary

accommodation   the norms that develop between ethnic groups regarding the ways they will interact with one another; the second stage of Park's race relations cycle

achieved status   position that is attained through individual efforts, such as one's education or occupation

advanced industrial societies   term used to refer to societies that have very productive extractive and manufacturing sectors and a growing service sector and a complex and productive communications industry

age-based norms   expectations and rules about behaviors that should occur at various ages

age effect   phenomenon whereby age-based norms influence our behaviors and actions

age/sex structure   see *population structure*

agent of socialization   the person or group that provides information about social roles

aggregates   large groups of people who actually have no relationship to one another except that they might happen to be in the same place at the same time

agrarian societies   societies with intense agricultural production made possible by the plow

alienation   term, first used by Marx, that refers to the separation of workers from the product or result of their work, which can result in feelings of powerlessness

anomie   Durkheim's term describing a situation of uncertainty over norms, or normlessness

anticipatory socialization   the process by which we prepare ourselves for future roles through thinking about and rehearsing the actions, emotions, and skills that may be involved in these new roles

ascribed status   position that is attained through circumstances of birth and that cannot be changed, such as one's race-ethnicity or gender

assimilation   the merger of two ethnic groups; the final stage of Park's race relations cycle

attributes   individual characteristics of persons or things; the values of variables

back-channel feedback   verbal and nonverbal conversational techniques used to let others know whether their spoken messages are being understood

backstage Goffman's term for the setting, or frame, in which impression management is not needed; contrast to *front-stage*

base term used by Marx and Marxian scholars to refer to the economy and the dynamics surrounding the means of production, which he saw as the basis of social structure

birth cohort people who are born in the same year; see also *cohort*

birth rates measures of the number of births in a society relative to the total number of people in the society

body language physical movements, postures, and gestures

bounded rationality term used to refer to the ways in which decisions made within organizations are constrained, or bounded, by the structure, technology, and culture of the organization

bourgeoisie term used by Marx and Marxian scholars to describe capitalists, people who own the factories and mills; contrast to *proletariat*

bureaucracies highly structured and formalized organizations that are governed by laws and rules

capitalism the free-enterprise economic system in which private individuals or corporations develop, own, and control business enterprises; contrast to *socialism*

capitalists people who own large industries, who continually reinvest the profits from these industries to increase their wealth, and whose power and social standing come from their control of capital (money and other tangible resources)

career the sequence of roles that we enact during our lifetimes

cartel organization of nations that produce similar products and cooperate to limit competition among one another and jointly set prices for their products

case study a research technique that involves an in-depth look at one case, such as one person, one group, or one organization

caste system a system of stratification like that traditionally found in India, whereby the family into which one is born determines one's social status and the types of occupations one can hold

causal model graphic device that illustrates sociological theories involving causal relationships between variables

causal relationship an association between variables in which one influences or causes the other

census a complete count of a group of people

census tracts small, relatively permanent areas within cities and towns that are designed by the Census Bureau to be fairly homogeneous in terms of population, economic situation, and living conditions and that represent, as much as possible, a city's neighborhood boundaries

charismatic authority Weber's term for authority that is based on the personal qualities of a particular individual

church a religious group that accepts the social environment in which it exists; compare with *sect*

class consciousness Marx's term for people in a social class who are aware of their common interests and concerns and of the fact that these interests conflict with those of another class group

closed-ended questions survey questions that give respondents only certain possible options from which to choose their answers

cognitive development emerging ability to understand and interpret the world

cognitive theory of social movements a theory of social movements that focuses on individuals' ideas and cognitive understandings and the ways in which these both prompt involvement in social movements and are shaped and developed by the actions of social movements

cohabitation the practice of men and women living together without marriage

cohort group of individuals who have some type of characteristic in common, such as the year in which they were born

cohort effect the common influence on attitudes and behaviors stemming from similar experiences based on cohort membership

collective behavior actions of groups of people, some of whom may not even know each other, that are often unexpected, spontaneous, and may involve unusual or atypical behaviors

collective consciousness Durkheim's term describing the common beliefs, values, and norms of people within a society

collective identity joint agreements about the nature of the group, who belongs and who doesn't belong, and what the group should and shouldn't do

command economy an economic system in which the political elite determine how goods and services will be produced, distributed, and consumed and in which workers' labor is coerced rather than freely hired

common schools schools held "in common," that is, for all children to attend

community a group of people who have frequent face-to-face interactions and common values and interests, relatively enduring ties, and a sense of personal closeness to one another

comparative research research method that involves the comparison of data from a variety of groups or settings, such as nations or historical eras

compensating differentials theory that employers compensate their employees for working in undesirable working conditions by paying them more

competition the inevitable conflicts between new and entrenched ethnic groups; the first stage in Park's race relations cycle

computer simulation the use of computer programs to simulate, or model, selected characteristics of real life; can be used by sociologists to conduct hypothetical experiments

concentric-zone model model of urban growth suggesting that cities expand outward from the central business district in a series of concentric circles or zones

concrete operational stage Piaget's term for the developmental stage, from about ages seven to eleven, in which children become able to handle logical operations

conglomerate corporate structure that includes many different firms specializing in entirely different types of products

conjugal family system a kinship system that emphasizes marriage ties over blood ties

consanguine family system a kinship system that emphasizes blood ties over marriage ties

conscience an internal set of ethical and moral principles that guide actions

construction industry industry that uses raw materials to build homes, offices, roads, and other structures

contact hypothesis hypothesis suggesting that individual discrimination and prejudice toward members of a minority group will diminish once members of the two groups have direct interpersonal contact

content analysis research method that involves the study of written documents or other types of recorded materials such as pictures, movies, or television shows

contingency approach theory of organizational leadership that suggests effective leadership traits and behaviors vary depending on a particular situation

control theory theory emphasizing that one's connections to others within society are the major influence on one's desire to conform to society's norms

control variable variable that is added to an analysis to see if it affects the relationship between an independent and a dependent variable

convergence theory theory that most countries around the world are moving toward the establishment of a mixed economic system

conversation analysis research technique involving close examination of patterns in verbal interactions

core nations term used by world systems theorists to refer to the most powerful, dominant countries in the global economy

corporate actors a unified group, such as an organization or nation, that engages in social actions

correlation the ways in which two variables may be related to each other in a predictable pattern

crime actions that a society explicitly prohibits and that are sanctioned through official means

crime rate a calculation of how frequently crime occurs for every 100,000 people within a population

cult   a religion that is in its beginning stages and is new to a society; the same as a new religious movement

cultural capital   Bourdieu's term describing how people behave, dress, and talk and how these manners and styles differentiate those in one class group from those in another

cultural diffusion   process whereby elements of one culture or subculture spread from one society or culture to another

cultural elements   basic aspects or characteristics of a culture, both material and nonmaterial

cultural gatekeepers   powerful organizations and individuals that control the entrance of new cultural elements into a society

cultural identity   the way in which people come to see themselves as belonging to or being part of a culture or cultural group

cultural innovation   process whereby new cultural elements are created

cultural lag   period of delay when one part of a society changes and other parts of a society have not yet readjusted

cultural materialism   theoretical perspective which suggests that societies develop differently primarily because they exist in different environments and thus tap different resources or materials

cultural universals   characteristics of a society that appear in all or virtually all cultures

culture   a common way of life; the complex pattern of living that humans develop and pass on from generation to generation

curvilinear   a relationship between two variables that is not well represented by a straight line and is better represented by some type of curved line

data   factual information that is used as the basis for making decisions and drawing conclusions; the plural form of the word datum

deductive reasoning   logical process of reasoning that moves from general theories or ideas to specific hypotheses or expectations; the opposite of inductive reasoning

deforestation   the disappearance of forested land, usually caused by overcutting of trees

democracy   government by the people, especially majority rule

democratic ideology   the belief that the state belongs to the people and should serve the interests of all citizens

democratic organization   an organization that is responsive to the desires of its members and does not degenerate into an oligarchy

democratic socialism   a mixture of capitalism and socialism whereby the government owns and operates some aspects of the economy and private industry owns and operates others; also called *welfare socialism*

demographic momentum   the growth of a society that occurs simply because of the people, or demographic characteristics, that are already there

demographic (population) processes   the ways in which populations change in composition and size

demographic transition theory   theory that summarizes the changes, or transitions, in population that societies experience as they move from the high fertility and mortality typical of preindustrial societies to the low fertility and mortality rates seen in highly industrialized societies

demography   the scientific study of populations and their effects

dependency ratio   The ratio of the dependent-age population, considered to be those younger than age 15 or older than age 64, to those of the traditional working ages (ages 15–64), used as a measure of how many people the average working person has to support.

dependent variable   variable that is said to be influenced or caused by another variable (the independent variable)

desertification   an increase in desert land, usually caused by overgrazing

developing countries   countries that are in the process of industrializing and that are in the middle of the global stratification system

deviance   behaviors that violate social norms

differential association theory   theory that focuses on how people learn deviant roles from their association or relationships with others engaged in deviant activities

diffuse status characteristics   status characteristics that go beyond specific task-related skills and reflect more general statuses, such as occupation, education, gender, age, or race-ethnicity

discrimination   differential treatment accorded to a group of people based solely on ascribed characteristics such as race-ethnicity

distributive justice   see *norm of equity*

diversification   form of corporate merger that results in a corporation specializing in many different aspects of one industry

divorce rate   the number of divorces in a year relative to every 1,000 existing marriages

domination   form of power in which one party controls the behavior of others through sanctions; compare with *influence*

dramaturgical theory   theory derived from the work of Goffman that uses the metaphor of a drama to explain how individuals play social roles and thus produce social structure

dysfunctions   negative consequences of a structure of society for the whole of society

ecological worldview   set of beliefs suggesting that new technology may create as many problems as it solves, that the earth's resources are limited, that the world's population has reached the earth's carrying capacity, and that population growth must be curbed

economic capital   income-producing wealth

economic exchanges   economic transactions; the transfer of goods and services in return for some type of valued item

economy   social institution that includes all the norms, organizations, roles, and activities involved in the production, distribution, and consumption of goods and services

education   social institution that includes the groups, organizations, norms, roles, and statuses associated with a society's transmission of knowledge and skills to its members

educational attainment   the number of years of education a person has

elites   powerful people who are able to influence the political process

elitism   view that political power and influence are dominated by a small handful of people who are relatively unified and form a comparatively small, tight-knit social network

emergent norms   norms that develop or emerge from a setting, rather than already being part of the well-known, and specified expectations established within a culture

empirical   based on experience and observation rather than pre-existing ideas

endogamy   marriage rule requiring people to select partners from within their own tribe, community, social class, or racial-ethnic or other such group

epidemiology   literally the study of epidemics, widespread outbreaks of communicable disease; today it also includes the study of other conditions that threaten health

ethnic enclave   area of a community in which there is a great deal of entrepreneurial business activity by members of a particular racial-ethnic group

ethnic group   group of people who have a common geographical origin and biological heritage and who share cultural elements, such as language; traditions, values, and symbols; religious beliefs; and aspects of everyday life, such as food preferences

ethnic identity   feeling of belonging specifically to an ethnic group

ethnocentrism   the belief that one's own culture or way of life is superior to that of others

evaluation research   research designed to examine the effectiveness of social change or intervention projects

exchange theory   Homans' theory that social action is an ongoing interchange of activity between rational individuals who decide whether they will perform a given action based on its relative rewards or costs

exclusion   term describing attempts to entirely remove lower-paid groups from a labor market, as, for example, through restrictive immigration laws

exogamy   marriage rule requiring people to select partners from outside their own tribe, community, social class, or racial-ethnic or other such group

expectation states   beliefs about what others within a group can do

expectation states theory   theory that group members' beliefs about what other members can do influence group interactions and the power structure within groups

experiment   research method that uses control groups and experimental groups to assess whether a causal relationship exists between an independent and a dependent variable

experimental variable   see *independent variable*

expressive leadership   leadership that deals with interpersonal relationships within a group or organization

extended family   family that includes relatives besides parents and children; contrast to *nuclear family*

external area   term used by world systems theorists to refer to countries that are largely excluded from the global economy and that are much poorer than other countries

external social control   attempts by others to control one's behavior

extractive industries   industries involved in the removal of raw materials from the environment, such as agriculture, fishing, forestry, and mining

falsification   the logic that underlies the testing of hypotheses; we can never prove that a theory is true, but can only say that it has either been falsified (shown to be untrue) or not yet been falsified

family development theory   theoretical perspective that looks at changes in family structure to understand how family life changes and how these changes affect individuals

family of orientation   the family in which one grows up

family of procreation   the family in which one lives as an adult and in which one has children (or procreates)

family structure   the combination of statuses within a family, particularly the number and type of statuses that are included

feminism   an ideology that directly challenges gender stratification and male dominance and promotes the development of a society in which men and women have equality in all areas of life

feminist movement   a social movement designed to promote the interests of women, much as the civil rights movement promoted racial-ethnic equality

fertility rates   measures of the number of births in a society relative to the total number of women of childbearing age

field experiment   experiment that takes place in a real-life setting as opposed to a laboratory

field research   research method in which a researcher directly observes behaviors or other phenomena in their natural setting

firm   the parent company or organization of work establishments

folklore   myths and stories that are passed from one generation to the next within a culture

folkways   norms that govern the customary way of doing everyday things

formal operations stage   Piaget's term for the final developmental stage in which children become capable of more complex and abstract thought

formal organizations   groups that have been deliberately created to accomplish certain goals

formal task groups   groups with specific goals, established norms, and recognized membership

frame   term used by dramaturgical theorists to refer to the setting in which interaction occurs and the norms that apply to a given situation

framing   the way social movements present and shape the ideas that underlie social movements

free riders   individuals who benefit from the public goods of a group or society even if they don't take the time or energy to participate

front-stage   Goffman's term for the setting, or frame, in which behaviors are designed to impress or influence others and in which impression management is important; contrast to *backstage*

function   the part a structure plays in maintaining or altering the society

game stage   Mead's term for the stage in which children begin to understand how different roles go together and become able to take the role of the other

Gemeinschaft Tönnies's term describing relationships that might appear in small, close-knit communities in which people are involved in social networks with relatives and long-time friends and neighbors, much like those that appear in primary groups

gender-based division of labor rules about what tasks members of each sex should perform

gender discrimination differential treatment of women because of their sex

gender identity one's gut-level belief that one is a male or a female

gender roles the norms and expectations associated with being male or female

gender schema cognitive framework that is used to organize information as relevant to one sex or the other

gender segregation the restriction of members of each sex group to different statuses and roles

gender segregation of the labor force the gender-based division of labor in the occupational world, or the phenomenon of men and women holding very different jobs; also called occupational gender segregation

gender socialization learning to see oneself as a male or female and learning the roles and expectations associated with that sex group

gender stratification the organization of society in a way that results in members of one sex group having more access to wealth, prestige, and power than members of the other sex group

general fertility rate the number of births per every 1,000 women of childbearing age in a given year

general strain theory theory of deviance that expands upon Merton's original formulation and suggests that the sources of structural strain can be much broader than Merton suggested and that the response which individuals choose is influenced by their perceptions of the fairness of the situation and the nature of their social networks

generalized other Mead's term for the conception people have of the expectations and norms that others generally hold; the basis of the "me"

Gesellschaft Tönnies's term describing relationships that come about through formal organizations and economic relationships rather than

kinship or friendship, similar to those that appear in secondary groups

ghetto highly segregated neighborhood populated primarily by people of one racial-ethnic heritage

Gini index a measure of income inequality within societies that can vary from zero, indicating that everyone receives an equal share of the country's income, to 1, indicating that all of the income is received by just one person

global economy economic exchanges and markets that include the entire world

globalization the process by which all areas of the world are becoming interdependent and linked with one another

global stratification system differences between nations that result in some societies having more power, prestige, and property than others

goods-producing industries industries involved in either the extraction, manufacture, or construction of products

grassroots organizations social movement groups that develop "from the ground up," away from major political centers

gross national product the value of all goods and services that a nation produces; GNP

group in sociological terms, two or more people who regularly and consciously interact with each other through engaging in some common activity and having some relatively stable social relationship

group boundaries patterns of behavior that define who does and who does not belong to a given group

groupthink process by which members of a group become so oriented to maintaining the cohesiveness of the group that they ignore or suppress information that may be critical of their decisions

group threat the extent to which members of a dominant group perceive that minority groups threaten their well-being; used as an explanation of variations in prejudice from one area or time period to another

growth rate a measure of how many people are added to a population each year divided by the number of people who are already in the population

hegemony   a hidden but pervasive power involving such extreme domination of social life that we seldom recognize it or question its legitimacy

hierarchy of identities   ordering of identities such that some identities are more important than others

historical–comparative research   research that uses historical data to compare two or more societies

historical research   research method that involves the examination of data from the past, often written artifacts and records

horizontal differentiation   the extent to which the work of an organization is divided up among different units or subgroups; part of organizational complexity

horticultural societies   societies that are based on a gardening economy

human capital   resources that we have as individuals and workers, such as our education, skills, and work experience

human capital theory   theory which suggests that different occupations and incomes reflect different amounts of human capital

hunting and gathering societies   societies that obtain their food through hunting animals and gathering plants and that, compared with other types of societies, use the simplest, most primitive tools and work techniques

hypersegregation   extreme isolation from other areas of the community

hypotheses   statements about the expected relationship between two or more variables; often derived from theories

ideal types   concepts or descriptions of phenomena that may not exist in a pure form in the real world but that define basic aspects of a given situation

ideologies   complex and involved cultural belief systems

imperialism   practice whereby one powerful country forcibly acquires territories in other areas of the world, creating vast colonial empires

impression management   Goffman's term describing how individuals may manipulate the impression or view that others have of them and give out cues to guide interactions in a particular direction

income   how much money a person receives in a given time, such as $20,000 a year

independent variable   variable that is said to cause or influence another variable (the dependent variable); also called the *experimental variable* in experiments

Index crimes   set of eight serious crimes used in the Uniform Crime Reports as the basis to calculate crime rates; see also *property crimes* and *violent crimes*

index of dissimilarity   a measure of segregation that is calculated from data on the characteristics of people living in different census tracts; also sometimes called a segregation index

individual discrimination   actions of individuals or small groups involving both discrimination and prejudice

inductive reasoning   logical process of reasoning that moves from specific ideas and observations to more general hypotheses and theories; the opposite of *deductive reasoning*

industrialized countries   countries that have a highly industrialized economy and that are at the top of the global stratification system

industrial worldview   the set of beliefs involving the notion that the earth's resources are valuable because they provide for humans' needs

industry   branches or areas of economic activity

infant mortality rate   the number of babies in 1000 who are expected to die before the age of one in a given year

influence   form of power whereby providing information and knowledge leads others to take different actions; compare with *domination*

informal groups   groups with no specified goals or formalized norms and no set membership

informal structure   social networks within an organization that exist alongside the organization's formal structure

institutional norms   norms that prescribe appropriate structures and behaviors of organizations and other aspects of social institutions

instrumental leadership   leadership that deals with the tasks or work of an organization

interaction effect   a pattern of influence in which the influence of one variable changes depending upon the status of another variable

interest groups   political organizations that concentrate their activities on specific policy issues or concerns

intergenerational mobility   holding a different occupational or social class position than one's parents

intergenerational succession   holding the same occupational or social class position as one's parents

internal migration   migration within national borders

international migration   migration across national borders

internalization   acceptance of the norms and behaviors of the group; the development of a conscience

internal labor market   the series of jobs that people may hold within an organization throughout their work careers

internal social control   control over one's behavior that is based on internalized standards; self-control

intervening variable   variable that comes between a dependent variable and an independent variable in a causal relationship

kin   group of relatives beyond the nuclear family

labeling theory   theory which suggests that definitions of deviant behavior develop from social interactions and that the key element in becoming deviant is how others respond to peoples' behavior, rather than how they actually behave

labor force   people employed within a society and, sometimes, those actively seeking work; work force

labor market   set of jobs within a society

labor market segmentation theory   theory that the labor market is divided into primary and secondary sectors and that women and members of racial-ethnic minorities earn less than others because they more often hold jobs in the secondary sector

labor unions   organizations of workers who have joined together to collectively promote their interests as workers

laissez-faire capitalism   capitalistic system in which the government has no control or influence over the operation of business

land reform   a policy that calls for redistributing very large parcels of land, many of which are now owned by extremely wealthy landowners or multinational corporations, to landless peasants

latent functions   functions that are less obvious and often unintended and that generally are unnoted by the people involved; contrast to *manifest functions*

laws   norms that are codified or written down; may be either folkways or mores

least developed countries   countries that are the least industrialized and that are at the bottom of the global stratification system

legal authority   Weber's term for authority based on written rules

life expectancy   the number of years a person is expected to live

literacy   the ability to read and write

longitudinal study   study which involves data that have been collected at different times

macrolevel   theories and analyses that deal with relatively broad areas of society rather than with individuals

mainstreaming   the practice of integrating special education students into the general curriculum and school day

male dominance   cultural beliefs that give greater value and prestige to men and to their roles and activities

manifest functions   functions that are easily seen and obvious; contrast to *latent functions*

manufacturing industries   industries that process raw materials into more usable forms

market   process of buying and selling; the way in which values are established for goods and services

mass schooling   elementary-level schooling for all children; common schools

material culture   the physical objects that are distinctive to a group of people, such as their food, clothes, houses, and hairstyles

mean   statistical average; computed by simply adding up all values and dividing by the number of cases

means of production   term used by Marx and Marxian scholars to refer to the way in which people produce their living, such as by farming or manufacturing or hunting and gathering

measure   the way in which concepts involved in a theory are translated into actual data

mechanical solidarity   Durkheim's term describing a society in which there is little differentiation, with people performing similar tasks, sharing similar responsibilities, and having similar behaviors; contrast to *organic solidarity*

median   the midway point in a distribution; the point at which 50 percent of the cases are larger and 50 percent are smaller

megalopolis   literally, "great city"; a term used to designate a vast area of land with a dense population

megamergers   form of corporate merger common since the 1980s that involves the union of some of the largest companies in the economy

mesolevel   theories and analyses that deal with social groups and organizations, such as individual families, classrooms, and work groups, rather than with very broad areas of society or with individuals

meta-analyses   the review and summary of a large number of studies of the same phenomenon to see what patterns emerge

methodology   the rules and procedures that guide research and help make it valid

metropolitan area (MA)   a region designated by the Census Bureau for statistical purposes that usually includes both the core area of a city with a large population and nearby communities

microlevel   theories and analyses that deal with relatively narrow aspects of social life, such as individuals' day-to-day activities and relations with other people

middle-range theories   theories that focus on relatively limited areas of the social world, as opposed to grand theories; often incorporate aspects of grand theories but are much more directed and applied toward specific research problems and can thus be more easily tested

migration rates   measures of the number of people who have moved into or out of an area relative to its total population

minority group   subculture that is subordinate to another group or groups within the society and that has less power, privilege, wealth, or prestige

mixed economic system   an economic system that blends elements of socialism and capitalism

mobility table   table of data used to illustrate patterns of intergenerational mobility and succession, typically involving the relationship between the occupations of parents and their offspring

modal   the most common or frequently occurring category or case

modernization theory   perspective suggesting that the majority of less developed countries will eventually industrialize in the manner of countries such as England and the United States

monogamy   the marriage of one man and one woman

monopoly   situation in which one company controls an entire market or industry

mores   norms that are vital to society, and violation of which is seen as morally offensive

mortality rates   measures of the number of deaths in a society relative to certain characteristics of the population

multinational corporations   business corporations that have outlets around the globe

multiple-nuclei model   model of urban growth suggesting that cities may have a number of different core areas (or nuclei), often influenced by the geographic and physical characteristics of an area

nation-state   political entity that unites large groups of people under agreed-upon laws and regulations

natural increase   the excess of births over deaths; the change in population size that comes from the difference in the number of births and deaths

neo-imperialism   system whereby one country dominates another through economic means rather than military or political control; said to characterize the post–World War II relationships of the United States with other countries

neo-Marxist   term used to describe recently developed theories that are in the Marxian tradition although they may depart from Marx's thought in certain ways

new religious movement a religion that is in its beginning stages and is new to a society; the same as a cult

new social movements recently developed social movements that tend to use tactics that have rarely been used before, especially the direct involvement of large numbers of citizens in political protest, and to address issues that have not been the focus of recent political debate, such as the quality of the environment, gender equality, alternative lifestyles, and human rights in developing countries

nodes see *social units*

nonmaterial culture the way of thinking of a group, including norms, values, ideology, folklore, and language

nonparticipant observation type of field research in which a researcher studies a group through observations without actually participating

nonverbal communication all the ways in which we send messages to others without words, including posture and movements, facial expressions, clothes and hairstyles, and manner of speaking

normative life stages periods during the life course when people are expected to perform certain activities

norm of equity belief that things should be "fair," that we and others should receive rewards that equal what we contribute to a relationship or interaction; also called *distributive justice*

norms cultural rules defining behavior that is expected or required within a group or situation; includes folkways, mores, and laws

nuclear family family group consisting of a mother, a father, and their children; contrast to *extended family*

occupational gender segregation see *gender segregation of the labor force*

occupational prestige scores scores given to occupations that indicate their relative prestige ranking

occupational socialization the process of learning and identifying with the norms and roles associated with a particular occupation

occupational status the occupation that one holds; often used in reference to the relative importance of one occupation over another

occupational structure occupations available within a society in a given time period

open-ended questions survey questions that allow respondents to give whatever responses they desire

organic solidarity Durkheim's term describing a society in which tasks, responsibilities, and behaviors are more highly differentiated, so that people are more dependent on one another; contrast to *mechanical solidarity*

organizational centralization the extent to which organizational power is centralized and decisions are made hierarchically

organizational complexity the extent to which the work of an organization is broken up and differentiated among various units; may be horizontal, vertical, and/or spatial

organizational culture the way of life, or culture, of an organization

organizational ecology theoretical perspective that looks at the relationship between organizations and their environments

organizational formalization the extent to which an organization uses written rules and procedures to control individuals within it

organizational leaders people within organizations who are able to influence the things others do and believe

organizational structure the way that the various parts of an organization are arranged; typically includes the elements of complexity, formalization, and centralization

organizational technology the work of organizations, including the skills and knowledge of workers in an organization, as well as the characteristics of the materials they work with and the machines they use

panel study a longitudinal study that includes information on the same people over a long period of time

paradigm a fundamental model or scheme that guides people's thinking about a particular subject

paradigm crisis the breakdown of an explanatory scheme, requiring a fundamental reorientation or rethinking of basic assumptions

paralanguage nonverbal aspects of speech, such as tone of voice and emphasis on words

participant observation type of field research in which a researcher studies a group or event while actually participating in it

peer review process by which research findings are evaluated by other experts in the field before publication to see if they meet the standard rules of research methodology

people of color people whose ancestors or who themselves came from non-European areas of the world and who can be identified through the color of their skin

per capita GNP the gross national product adjusted for the number of people who live in the country

period effect the influence of the norms and events of a particular historical time on individuals' lives

peripheral nations term used by world systems theorists to refer to countries that sell raw materials to core nations and that are at the edge of the global economy

play stage Mead's term for the developmental stage in which children "play at" or assume various roles one at a time

pluralism view that the political power structure involves a number of powerful groups and individuals, all of which can potentially influence the decision-making process

plutocracy government by the wealthy

political economy the interrelationship of economic and political processes

polity the social institution that includes all of the various ways that societies have developed to maintain social order and control; the political world

polyandry a system of marriage whereby the wife can have more than one husband; the opposite of *polygyny*

polygyny a system of plural marriage whereby the husband has more than one wife

population the entire group or set of cases that a researcher is interested in generalizing to; see also *sample*

population projections estimates of the size and composition of a population in future years

population pyramid diagram used by demographers to illustrate the population or age/sex structure of a society

population structure the age and sex composition of a society; the age/sex structure

poverty line the income level under which families are officially defined as being poor

poverty rate the percentage of the total population that lives in families with incomes under the official poverty level

power the ability of one social element, either a group or a person, to compel another social element to do what it wants

power structure the power differences within a group; the way that power is distributed in a society

PPE spiral term used to refer to the intertwining influences of poverty, population growth, and environmental degradation

prejudice preconceived hostile attitudes toward a group of people simply on the basis of their group membership

preoperational stage Piaget's term for the developmental stage, from about eighteen months to seven years of age, in which children learn to use symbols and communicate with language but don't yet have the mental flexibility to perform mental operations that involve complex relationships

prestige classes aggregates or clusters of people who possess similar characteristics, who are perceived by their fellow citizens as being similar, and who are accorded similar levels of respect and esteem

primary groups groups that include only a few people and that are characterized by intimate, face-to-face interaction

primary labor market "good" jobs with good wages and benefits, comfortable working conditions, job security, and chances for advancement; contrast to *secondary labor market*

probability sample sample that can be generalized to a larger group, typically chosen through some type of random selection process

proletariat term used by Marx and Marxian scholars to describe the workers; contrast to *bourgeoisie*

property crimes crimes that don't involve the use of violence or force against individual people, such as burglary, larceny-theft, motor vehicle theft, and arson

public goods necessities of group life that individuals cannot provide by themselves but must obtain through cooperation with others; common goods

qualitative data measures of data that cannot be assigned real numbers; contrast to *quantitative data*

quantitative data measures that may be assigned real, or meaningful, numbers—for example, income or age

racial-ethnic group subculture that can be distinguished on the basis of skin color and ethnic heritage

racial-ethnic stratification the organization of society such that people in some racial-ethnic groups have more property, power, or prestige than do people in other groups

random selection process that gives each member of a population an equal chance of being included in a sample

rational choice theory theory suggesting that people make decisions by balancing the costs and benefits involved and choosing the actions that provide the lowest costs and the greatest benefits

reference group group of people that we look to, or use for reference, in evaluating ourselves and our position in life

reflexive behavior the process of being able to think about one's own actions and mentally assume the role of other people

regional stratification differences within a country that result in some regions having more power, prestige, and property than others

relative deprivation a gap between people's expectations and the actual state of their lives

reliability the extent to which a measure yields the same results when used by different researchers on the same subject at different times

religion the social institution that deals with the area of life people regard as holy or sacred; it involves the statuses, roles, organizations, norms, and beliefs that are related to humans' relationship with the supernatural, including shared beliefs, ethical rules, rituals and ceremonies, and communities of people with common beliefs and standards

religious economy the various religious options or choices that are available within a society

replacement rate the fertility rate that is needed for a population to replace itself or to remain stable in size (net of migration)

replication repetition of an earlier study to see if the same results occur and if they hold in other settings

representative democracy a political system in which elected officials represent the citizens and make important governmental decisions

repression psychoanalytic term for the notion that people repress, or push out of conscious awareness, ideas that are uncomfortable or painful to think about

resource mobilization theory theory suggesting that social movements are successful to the extent that they can effectively mobilize or activate the resources needed to accomplish their goals

rite of passage ceremony or ritual that helps individuals deal with role transitions through the life course

rituals cultural ceremonies that often mark important life events

role conflict situation in which a person holds roles with incompatible norms or obligations

role theory perspective that social structure is created and maintained because people generally act in ways that conform to social roles

role transition moves, or transitions, from one role to another during the life course

rural development industrialization in rural areas

sample subset of a larger group or population; see also *probability sample*

sanctions social reactions to an individual's behavior, generally reflecting attempts to control the behavior; rewards and punishments

scheme a mental framework or set of rules to understand how the world works

schooling formal instruction by trained teachers, usually within schools

secondary analysis analysis of data that have already been gathered by other researchers

secondary deviance deviance that results from the process of being labeled as a deviant

secondary groups groups that involve more than just a few people and relatively distant social relations

secondary labor market "bad" jobs with low pay, few if any benefits, little job security, and few chances for advancement; contrast to *primary labor market*

sect a religious group that rejects the social environment in which it exists; compare with *church*

sector model model of urban growth suggesting that cities are often composed of sectors or segments, which extend out from the central business district

secularization process of transformation to an outlook on life based on science and reason rather than on faith and supernatural explanations

segmented occupational structure division of occupations into "good jobs," characterized by stable employment, benefits, and higher salaries, and "not-so-good jobs," which lack these characteristics

segregation the division or separation of neighborhoods in ways that lead to the inclusion of some groups and the exclusion of others

segregation index see *index of dissimilarity*

self one's view of oneself as a distinct person with a clear identity

self–concept the thoughts and feelings we have about ourselves

self–identity a set of categories used to define the self; the way we think about ourselves

semiperiphery term used by world systems theorists to refer to nations that fall between the core and periphery

sensorimotor stage Piaget's term for the developmental stage, from birth to about eighteen months, in which infants learn through their senses and their movements

serial monogamy a pattern of marriage in which a person has several spouses over a lifetime, but only one at a time

service industries global term used to describe the entire set of non-goods-producing industries; also used to refer to a specific set of industries that provide various nonmaterial things people want or need, such as medical care, education, social welfare, and entertainment

significant others people with whom you interact and who are emotionally important to you

situated self the subset of self-concepts or self-identities that apply to one's self-view and behaviors in a given situation

social action day-to-day decisions and actions of individuals within the social world; social actions both influence and are patterned and influenced by social structure

social capital resources or benefits people gain from their social networks

social change the way in which societies and cultures alter over time

social class groups of people who occupy a similar level in the stratification system

social control efforts to help ensure conformity to norms

social distance the types of social ties or interactions people are willing to establish with others

social ecology the study of the ways in which people interact with their physical and geographic environment

social institutions the complex sets of statuses, roles, organizations, norms, and beliefs that meet people's basic needs within a society

socialism an economic system that involves public rather than private ownership of the means of production; contrast to *capitalism*

socialization the way in which we develop, through interactions with others, the ability to relate to other people and to play a part in society

social learning perspective theory of socialization suggesting that we learn behaviors by being rewarded for those that conform to norms and by imitating those that we think conform to roles we want to fill

social mobility movement between social class groups

social movement adherents organizations and individuals that believe in the goals of a social movement

social movement constituents organizations and individuals that provide resources for a social movement

social movement industry the set of social movement organizations that are all working within the same social movement

social movement organizations formal organizations that work toward the goals of a social movement

social movements organized and concerted efforts to promote social change by groups of individuals

social movement sector the entire range of activities that are aimed at changing the social structure

social network patterns or webs of social relationships; the linkages between individuals formed by social interactions

social role expectations, obligations, and norms that are associated with a particular position in a social network

social status positions that individuals occupy within the social structure

social stratification the organization of society in a way that results in some people having more and some people having less; divisions in a society based on social class

social structure relatively stable patterns that underlie social life; the ways in which people and groups are related to each other, and the characteristics of groups that influence our behavior

social ties relationships between units of a social network

social units elements of a social network; also called *nodes*

society a group of people who live within a bounded territory and who share a common way of life

sociological imagination Mills's term describing the ability to discern patterns in social events and view personal experiences in the light of these patterns

sociology the science of society; the scientific study of the social world and social interactions

spatial dispersion the extent to which the various units of an organization are spread out in different locations; an element of organizational complexity

specific status characteristics status characteristics that are closely related to the given tasks of a group; contrast to *diffuse status characteristics*

split labor market the division of the labor market into three class groups: business owners and employers, higher-paid workers, and lower-paid workers

spurious correlation correlation between two variables that only occurs because of the influence of a third variable

state the organized monopoly or control of the use of force in a society; synonymous with government or the polity

statistical discrimination theory that employers have beliefs, or stereotypes, about the relative stability and productivity of men and women workers, which they rely on in hiring employees; when these average characteristics don't apply to individual women, statistical discrimination is said to have occurred

status term used in the Weberian tradition to designate one dimension of stratification, that involving communities or social networks of people with similar lifestyles and viewpoints; synonymous with *prestige*

status attainment model theoretical model that describes the way variables, such as family background, individual motivations and ability, and interactions with significant others, influence the educational levels and occupations and incomes we have in adulthood

status characteristics the social statuses that people hold and the evaluations and beliefs (characteristics) that are attached to these statuses

status generalization the process by which diffuse status characteristics influence group interactions

strain theory Merton's theory that deviant behavior results when individuals accept culturally defined goals but do not have the institutionalized means of attaining these goals

stratification system see *social stratification*

structural discrimination discrimination that results from the normal and usual functioning of the society (the social structure) rather than from prejudice or from laws and norms that promote segregation or exclusion

structural functionalism sociological theory that tries to account for the nature of social order and the relationship between different parts of society by noting the ways in which these parts or structures function to maintain the entire society

structural mobility intergenerational mobility resulting from changes in the occupational structure

subcultural theory of deviance theory suggesting that deviant behavior can reflect conformity to the norms of subcultures that support deviant lifestyles

subculture group or culture that exists within a broader culture

suburb a city or town outside a city's boundaries but either adjacent to the city or within commuting distance

superstructure term used by Marx and Marxian scholars to refer to the areas of social life that they believe are influenced by the economy, such as religious beliefs, family relations, and political life

survey research method of data gathering that involves asking people questions, through either interviews or written questionnaires

sustainable development economic development that does not deplete or degrade natural resources.

symbolic interactionism theory that social interaction involves a constant process of presenting and interpreting symbols through thinking about what another person is trying to communicate through the use of symbols

symbols anything that people use to represent something else; for example, language uses the symbols of words to represent objects and ideas

synnomie the state of a society that has a strong collective consciousness and a high degree of cohesion; the opposite of *anomie*

target of socialization the person who is learning a social role, who is being socialized

technological innovations the development of new material elements of culture that help people deal with the problems of day-to-day life

technology the application of knowledge to solve problems

theories broad systems of ideas that help explain patterns in the social world

traditional authority Weber's term for authority that is accepted because it is an integral part of the social structure and it is impossible to conceive of any other way of being

trend analysis study that examines data collected at different time points but that uses a different sample each time

turn taking term used by conversation analysts to describe the conversation process whereby one person talks and then the other talks; results from both verbal and nonverbal cues

typology a classification of a group or phenomenon into discrete categories

unconscious psychoanalytic term for thoughts and impulses individuals may have that they aren't consciously aware of

unobtrusive research research method in which a researcher obtains data without directly talking to or watching people

upward mobility social mobility that involves movement to a social class position that is higher than one's parents occupied

urbanization the process of societal change that involves the movement of people from rural areas or small towns to metropolitan areas

validity the extent to which a measure actually represents the concept it is said to be measuring; when applied to a research design, indicates that we can trust the conclusions

values general standards about what is important to a group

variables logical groupings of attributes; literally, things that vary or have more than one value

vertical differentiation the number of supervisory levels in an organization; part of organizational complexity

vertical integration   form of corporate merger producing companies that control the entire extraction, manufacturing, and distribution process

violent crimes   crimes that involve the actual use or threat of force against people, including murder, forcible rape, robbery, and aggravated assault

vital statistics   data that record the "vital events" of life, such as births, deaths, marriages, divorces, and migration

waves of data collection   times of data collection in a panel study

wealth   assets resulting from the accumulation of income, such as houses, cars, real estate, and stocks and bonds

welfare capitalism   form of capitalism in which a broad system of laws protects the welfare of workers and consumers in their economic transactions

welfare socialism   see *democratic socialism*

white-collar crime   nonviolent crimes that generally involve fraud and deception and are committed in the workplace

white ethnics   people who identify themselves as belonging to a group with European origins, usually other than Great Britain

wholesale and retail trade industry   industry involving the sale of goods to stores and directly to individual consumers

work establishment   the actual place where someone works; compare with *firm*

work force   see *labor force*

world systems theory   perspective developed by Wallerstein to explain global stratification; suggests that a set of core, highly developed nations maintain their wealth by dominating and exploiting other nations, said to be on the periphery of the world economy

zero population growth   the idea that the world population will stop growing when birth rates equal death rates and when, on average, each woman has only two children